T0231077

COMPLEX NETWORKS

An Algorithmic Perspective

OTHER COMMUNICATIONS BOOKS FROM AUERBACH

Anonymous Communication Networks: Protecting Privacy on the Web
Kun Peng
ISBN 978-1-4398-8157-6

Case Studies in System of Systems, Enterprise Systems, and Complex Systems Engineering
Alex Gorod, Brian E. White, Vernon Ireland,
S. Jimmy Gandhi, and Brian Sauser (Editors)
ISBN 978-1-4665-0239-0

Cyber-Physical Systems: Integrated Computing and Engineering Design
Fei Hu
ISBN 978-1-4665-7700-8

Evolutionary Dynamics of Complex Communications Networks
Vasileios Karyotis, Eleni Stai, and Symeon Papavassiliou
ISBN 978-1-4665-1840-7

Fading and Interference Mitigation in Wireless Communications
Stefan Panic, Mihajlo Stefanovic, Jelena Anastasov, and Petar Spalevic
ISBN 978-1-4665-0841-5

Green Networking and Communications: ICT for Sustainability
Shafiullah Khan and Jaime Lloret Mauri (Editors)
ISBN 978-1-4665-6874-7

Intrusion Detection in Wireless Ad-Hoc Networks
Nabendu Chaki and Rituparna Chaki (Editors)
ISBN 978-1-4665-1565-9

Intrusion Detection Networks: A Key to Collaborative Security
Carol Fung and Raouf Boutaba
ISBN 978-1-4665-6412-1

Machine-to-Machine Communications: Architectures, Technology, Standards, and Applications
Vojislav B. Mišić and Jelena Mišić (Editors)
ISBN 978-1-4665-6123-6

MIMO Processing for 4G and Beyond: Fundamentals and Evolution
Mário Marques da Silva and Francisco A. Monteiro (Editors)
ISBN 978-1-4665-9807-2

Network Innovation through OpenFlow and SDN: Principles and Design
Fei Hu (Editor)
ISBN 978-1-4665-7209-6

Opportunistic Mobile Social Networks
Jie Wu and Yunsheng Wang (Editors)
ISBN 978-1-4665-9494-4

Physical Layer Security in Wireless Communications
Xiangyun Zhou, Lingyang Song, and Yan Zhang (Editors)
ISBN 978-1-4665-6700-9

SC-FDMA for Mobile Communications
Fathi E. Abd El-Samie, Faisal S. Al-kamali,
Azzam Y. Al-nahari, and Moawad I. Dessouky
ISBN 978-1-4665-1071-5

Security for Multihop Wireless Networks
Shafiullah Khan and Jaime Lloret Mauri (Editors)
ISBN 978-1-4665-7803-6

Self-Healing Systems and Wireless Networks Management
Junaid Ahsenali Chaudhry
ISBN 978-1-4665-5648-5

The State of the Art in Intrusion Prevention and Detection
Al-Sakib Khan Pathan (Editor)
ISBN 978-1-4822-0351-6

Wi-Fi Enabled Healthcare
Ali Youssef, Douglas McDonald II, Jon Linton,
Bob Zemke, and Aaron Earle
ISBN 978-1-4665-6040-6

Wireless Ad Hoc and Sensor Networks: Management, Performance, and Applications
Jing (Selina) He, Shouling Ji, Yingshu Li, and Yi Pan
ISBN 978-1-4665-5694-2

Wireless Sensor Networks: From Theory to Applications
Ibrahiem M. M. El Emary and S. Ramakrishnan (Editors)
ISBN 978-1-4665-1810-0

ZigBee® Network Protocols and Applications
Chonggang Wang, Tao Jiang, and Qian Zhang (Editors)
ISBN 978-1-4398-1601-1

AUERBACH PUBLICATIONS
www.auerbach-publications.com
To Order Call: 1-800-272-7737 • Fax: 1-800-374-3401
E-mail: orders@crcpress.com

COMPLEX NETWORKS

An Algorithmic Perspective

Kayhan Erciyes

CRC Press
Taylor & Francis Group
Boca Raton   London   New York

CRC Press is an imprint of the
Taylor & Francis Group, an **informa** business

CRC Press
Taylor & Francis Group
6000 Broken Sound Parkway NW, Suite 300
Boca Raton, FL 33487-2742

© 2015 by Taylor & Francis Group, LLC
CRC Press is an imprint of Taylor & Francis Group, an Informa business

No claim to original U.S. Government works

Printed on acid-free paper
Version Date: 20140916

International Standard Book Number-13: 978-1-4665-7166-2 (Hardback)

This book contains information obtained from authentic and highly regarded sources. Reasonable efforts have been made to publish reliable data and information, but the author and publisher cannot assume responsibility for the validity of all materials or the consequences of their use. The authors and publishers have attempted to trace the copyright holders of all material reproduced in this publication and apologize to copyright holders if permission to publish in this form has not been obtained. If any copyright material has not been acknowledged please write and let us know so we may rectify in any future reprint.

Except as permitted under U.S. Copyright Law, no part of this book may be reprinted, reproduced, transmitted, or utilized in any form by any electronic, mechanical, or other means, now known or hereafter invented, including photocopying, microfilming, and recording, or in any information storage or retrieval system, without written permission from the publishers.

For permission to photocopy or use material electronically from this work, please access www.copyright. com (http://www.copyright.com/) or contact the Copyright Clearance Center, Inc. (CCC), 222 Rosewood Drive, Danvers, MA 01923, 978-750-8400. CCC is a not-for-profit organization that provides licenses and registration for a variety of users. For organizations that have been granted a photocopy license by the CCC, a separate system of payment has been arranged.

Trademark Notice: Product or corporate names may be trademarks or registered trademarks, and are used only for identification and explanation without intent to infringe.

Library of Congress Cataloging-in-Publication Data

Erciyes, Kayhan.
    Complex networks : an algorithmic perspective / author, Kayhan Erciyes.
        pages cm
    Includes bibliographical references and index.
    ISBN 978-1-4665-7166-2 (hardback)
    1. System analysis--Mathematics. 2. Computational complexity. 3. Algorithms. I. Title.

T57.E73 2014
003'.72--dc23   2014028247

Visit the Taylor & Francis Web site at
http://www.taylorandfrancis.com

and the CRC Press Web site at
http://www.crcpress.com

Dedication

To my family and all lifelong learners.

Contents

SECTION II: ALGORITHMS   79

List of Figures

List of Tables

Preface

Recent technological advances in the last two decades have provided availability of enormous amounts of data about large networks consisting of hundreds and thousands of nodes. These so-called *complex networks* have non-trivial topological features and can vary from technological networks to social networks to biological networks. The study of complex networks, sometimes referred to as *network science*, has become a fundamental research area since then in various disciplines such as mathematics, statistics, computer science, physics and biology.

These seemingly unrelated networks experimentally have been shown to have common properties such as low average distance between their nodes, high local densities and degree distributions with few high degree nodes and many low degree nodes. Modeling and analysis of these networks based on experiments and evaluations has become an active and attractive area of research with many potential results. Graphs have been widely and successfully used to model computer networks, and it seems graph theory is a promising tool also for complex networks. Although there has been considerable amount of study and research on the modeling and analysis of complex networks, the algorithms for these networks are relatively less investigated.

Whether a complex network is man-made such as the Internet or not such as a protein interaction network, predicting its behavior is not a trivial task. Understanding the functionality of complex networks provides us with insight to predict their behavior and once we can estimate the behavior of a complex network based on its functionality, we may be able to control its functionality. For example, if we can understand the spreading pattern of an epidemic disease which in fact is a complex network, we can estimate where it will most likely spread and can then take precautions to stop it. In summary, controlling the functionality of a complex network is one of the fundamental reasons to study these networks.

As a first step in their study, we need to specify and classify the properties of complex networks. We can then use analytical tools to identify and analyze these properties to understand them better. As an example, a group of entities that make the complex network may be more closely related to each other than the rest of the network. These groups called the *clusters* may have important processing effects on the overall functionality of the network. If we can detect clusters in a social network for example, we can locate these intense regions of activity in that network after which we can investigate the role of these clusters in the functioning of the whole network. Detection of these properties may be visually possible in a small network of few tens of nodes but for a complex network of hundreds of thousands of nodes, we need analytical tools and computational methods. Properties of complex networks such as clustering depend on their topological properties and study of topological properties of these networks provides insight to their functioning.

This book is about specifying, classifying, designing and implementation of mostly sequential, and also parallel and distributed algorithms that can be used to analyze mostly the static properties of complex networks. Our aim has been to identify and describe a *repertoire* of algorithms that may be of use for *any* complex network. The starting point was to identify fundamental and mostly topological properties which are static in general and evaluation of these properties which requires efficient algorithms. The problems encountered are NP-hard in many cases and we need to rely on *approximation algorithms* where sub-optimal solutions in polynomial time can be found. Sometimes, using *heuristic algorithms* may be the only choice and extensive tests are needed to support that the algorithm works for most of the cases. *Parallel* algorithms aim at performance and provide efficiency for computation intensive tasks to be performed and we present several parallel algorithms. *Distributed algorithms* reach a decision by local information and are usually the only choice in computer networks. Design and implementation of parallel and distributed algorithms received little attention for complex networks in the past and are promising areas for potential research in these networks.

An important static topological property of a complex network is its *centrality* measure which shows the importance of a node or an edge in the network. *Clustering* or *community detection* is another fundamental topological complex network property and provides information about groups of nodes in the complex networks which have closer relations among them than the rest of the network. Discovery of *motifs* which are patterns occurring more than any other patterns, possibly indicating a basic function in the network is another important property of the complex networks. Evaluation of such measures using sequential, approximation, heuristic, parallel and distributed algorithms provides us with significant information about a particular network. We can then improve the modeling of the network, understand its function better and possibly predict the behavior of the network.

The style we have adopted is to keep everything as simple as possible, to be able to guide a beginning researcher or a student with virtually no background in the field of complex networks. The language used is mathematical rather than descriptive most of the time; however, a basic discrete mathematics and algorithms background at undergraduate level is sufficient to follow the material. Again, to aid the beginner in the field, most of the algorithms are provided in ready-to-be-executed form to test.

The book is divided into three parts. Part I provides the basic background in terms of the graph theory; algorithms and complexity and the specification of the parameters for the analysis of complex networks. In Part II, we provide a survey of important algorithms for the analysis of complex networks, starting with distance and centrality algorithms. We then describe algorithms to construct and detect special subgraphs in complex networks, which may be used for other tasks such as clustering. A survey of data and graph clustering algorithms is also presented and this part concludes by the description of the network motif discovery algorithms. Part III is about case studies of complex networks and we show the implementation of some of the algorithms we have described in real-life networks such as the protein interaction networks, the social networks and the computer networks.

I would first like to thank graduate students at Izmir University who were concurrently taking a related course at the time of the writing of this text and were presented part of the material. I would like to thank Esra Ruzgar and Can Ileri for their feedback and especially Vedat Kavalci for proofreading of several chapters. I would like to thank CRC Press publisher Rich O'Hanley who has always been very kind, supportive and encouraging. I also thank Stephanie Morkert who was prompt and ever willing to help in the editing process and Michele Dimont for final editing.

K. Erciyes
Izmir, Turkey

Chapter 1

Introduction

1.1 Overview

A network is a set of elements with links connecting these elements. A computer network for example, consists of computing nodes that are connected using communication links. Any network can be conveniently modeled by a *graph* structure $G(V,E)$ where V is a set of vertices representing the elements of the network, and E is a set of edges representing the links. Figure 1.1 displays a network with a vertex set $V = \{a,b,c,d,e,f,g,h\}$ and an edge set $E = \{(a,b),(b,h),(b,g),(b,f),(b,c),(c,f),(d,f),(g,f),(e,f)\}$.

A *complex network* is a network with non-trivial topological features. Common characteristics of complex networks are as follows :

- They are large, consisting of hundreds and thousands or more elements.

- Patterns of connections are neither regular nor purely random.

- They change dynamically.

- They form the backbones of complex systems.

Recent advances in technology, including increased computational power and size of storage space, enabled gathering of vast amounts of data from these networks, and hence analysis of complex networks has become possible in much more detail than previously. Complex networks, most of the time, are dynamic in nature and their topology evolves with time. A general consensus among researchers of complex networks is that their topology has a profound effect on their behaviour. Therefore, analysis of complex networks based on their topology provides us with data that can be used to estimate the functioning and growth of these networks. Graph theory and graph algorithms have been used extensively for the analysis of these networks but

1

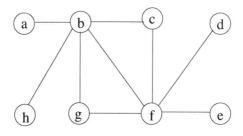

Figure 1.1: A graph representing a network

graph theory alone has not been adequate in understanding the behaviour of these complex systems.

1.2 Real-world Complex Networks

Real world networks can be categorized as technological networks, information networks, social networks, and biological networks [4]. We will describe these types of complex networks briefly in the following sections.

1.2.1 Technological Networks

Technological networks are man-made networks designed to distribute some resource or commodity. The Internet, the telephone network, electrical power grid, sensor networks, ad hoc wireless networks and road networks are the examples of technological networks.

The Internet connects different computer networks that are distributed world wide. An important property of the Internet is the heterogeneity of its components. The computers connected over the Internet, their operating systems and their physical connections to each other vary extensively. Packets are the main units of data transferred over the Internet and *routing* is the process of transferring a packet from a source computer to a destination computer using lowest cost paths by *packet switching*. A router is a device that performs routing of packets. The Internet can be considered as a graph with routers as nodes of this graph at *Internet router* level, or the nodes being autonomous systems that are governed by the same set of protocols at *autonomous system* level. Protocols are mainly software modules running at the nodes of the Internet to transfer data packets reliably. The Internet uses a number of protocols structured as hierarchical layers, each of which is responsible for certain tasks. For example, the physical layer is the lowest layer and performs tasks such as signalling and bit synchronization.

The telephone network on the other hand, is used to transfer telephone calls between the subscribers. It is one of the oldest networks in the world and mainly uses *circuit switching* where a certain transmission medium is allocated between the callers for the duration of the call. Modern telephone networks may employ packet

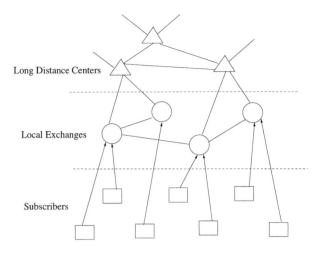

Figure 1.2: The telephone network

switching partly to transfer calls. The telephone network uses a three level hierarchical structure consisting of long distance offices at the top, local exchanges in the middle and subscribers at the lowest level as shown in Figure 1.2 where each subscriber is connected to local exchanges and local exchanges are connected to long distance offices which are connected among themselves.

1.2.2 Information Networks

These networks are characterized by the flow of information between their nodes. Examples of information networks are the *citation networks* and the *World Wide Web* or simply the Web. Academic papers are cited in citation networks and pages are transferred in the Web. The Web consists of a huge number of sites, each containing a collection of documents and links to other sites using *hyperlinks*. Each site is identified by a unique *domain name* which can be accessed using the Web tools. The Web can be represented by a directed graph where web pages are the vertices of this graph and the hyperlinks are shown by the directed edges from a Web page to the page referenced by the hyperlink. The citation networks are formed by the authors of articles that cite other articles. In this case, the edges of the graph are directed from the citing articles to the reference articles as shown in Figure 1.3 where articles are shown in a time frame from the newest at the top to the oldest at the bottom and there are no cycles as authors of the articles can only cite articles written before them.

1.2.3 Social Networks

A social network is a set of people or or groups of people with some pattern of contact or interactions among them. Interactions may be in the form of friendship,

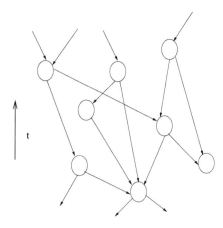

Figure 1.3: An example citation network

business relations or some other form of relationship. Data from social networks can be directly obtained by questionnaires which limit the amount of data that can be obtained. For example, to build the friendship network in a school, students may be asked who their best friends are. The relationship in this case can be represented by a directed graph as it is not symmetric. A person a can consider another person b as one of her best friends but b may not think a is one of her best friends.

Collaboration networks are the types of social networks where individuals belonging to a group are linked. In the film actors social network for example, the actors that have appeared in the same film are connected. Similarly, in the coauthorship network, academicians are connected if they have coauthored a paper.

Social networks tend to have some interesting properties. Firstly, they exhibit *small world property* which means the diameter of a social network in general is small. Sociologist Milgram [7] performed a test where a number of mails were sent by people who did not know the recipients, but they sent these mails to people who may have known someone in connection with the destination addresses. In the end, a quarter of the mails reached their destinations using a maximum of 6 hops. Another significant characteristic of these networks is that they have a small number of highly connected nodes called *hubs* and a large number of nodes with low degree. They also exhibit *assortative mixing* where highly connected nodes are usually linked to each other. In a friendship network, this property manifests itself as the popular people being friends of each other.

1.2.4   *Biological Networks*

Biological networks vary from genetic regularity networks to protein interaction networks and to ecological networks. The building block of an organism is the cell and the ingredients of a cell include desoxyribonucleic acid (DNA), ribonucleic acid

(RNA), proteins and metabolites at molecular level. Cells make up *tissues* which constitute the *organ* of an organism. The *ecosystem* is formed by many different organisms. At each level of life, the relation and interaction among the elements can be represented by networks. The biological entities are represented by the nodes and their interaction by the edges of the network.

The fundamental biological networks at molecular level are the *protein interaction networks*, *gene regulation networks*, *metabolic networks* and *signal transduction networks*. The DNA is enclosed within nucleus of the cell of an organism and consists of a helix structure with two backbones which consist of pairs of nucleotide bases: adeline (A), cytosine (C), guanine (G) and thymine (T). The nucleotide A only pairs with T, and C only pairs with G.

DNA has the static information in the cell as a sequence of these four nucleotides which make up the genes and proteins are involved in dynamic activity within the cell using DNA information. Information transfer from DNA to proteins is called *gene expression* which is further divided into *transcription* and *translation* processes. During the transcription process, a strand of DNA is read and the corresponding RNA blueprint is formed. In the translation process, amino acid sequences are formed from the RNA by the ribosomes. RNA nucleotide acid sequence is read in triplets called *codons* to encode 20 *amino acids*. Proteins consist of a number of amino acids, and this sequence along with their 3D structure such as loops and helices affects their function. Proteins interact with other proteins in the cell to perform variety of functions such as to stabilize the structure of the cell and regulate the process of transcription. These interactions can be conveniently modeled by *protein interaction networks* (PINs), sometimes referred as *protein-protein-interaction* (PPI) networks, where the nodes of the network graph are the proteins and the existence of edges between the proteins indicate interactions.

Metabolites in general are small molecules such as amino acids and glucose in the cell. Metabolic pathway is a sequence of biochemical reactions to perform a specific metabolic function. Metabolic networks have metabolites as their nodes which are transformed to each other using *enzymes*. The metabolic network of a cell shows all the material processing in the cell to generate energy and synthesize its vital components.

Ecological networks on the other hand, show the interaction between various organisms on a much larger scale. *Food web* is an example of an ecological network which can be modeled by a digraph where edges show predator-prey relations between the nodes. *Phylogenetic networks* on the other hand, display the evolutionary relations between organisms.

1.3   Topological Properties of Complex Networks

Complex networks have certain common topological attributes which are frequently observed in their experimental evaluations. They exhibit for example, small world property which means the average distance between any two nodes of a complex network is small. They are also *scale-free* in general, where they include only few nodes with high degrees and the rest of the nodes usually have low degrees.

It is in general not possible to visualize a complex network as a whole which may consist of hundreds of thousands of nodes. We need methods other than visualization to determine the structure of these networks.

Centrality of a node or an edge is a measure of its importance in the network. A simple way to find the centrality of a node is to determine its degree which is the number of edges incident to it or the number of its neighbors. For example, a person with a high number of connections in a social network may show that person is influential in that network. *Clusters* or *communities* in complex networks consist of nodes that are more closely related to each other than the rest of the network and detecting such group of nodes in a complex network is an important step in understanding its structure.

A *motif* in a complex network graph is a small graph structure that repeats itself significantly more than other structures. Motifs are the building structures of biological networks and detecting and counting motifs in these networks may provide insight to the structure and functioning of these networks. A complex network such as a biological network evolves with time to produce two or more networks that are similar but not equal. In this case, we are interested to find the similar structures in two or more networks, to understand their common origins. *Network alignment* is the process of evaluating similarity between networks that appear and function different.

1.4 Algorithmic Challenges

In order to be able to understand the behaviour of complex networks, we need to analyze their static and dynamic properties in detail. Efficient algorithms have an important role in the analysis of aforementioned properties.

A fundamental issue in this area is the identification of the class of algorithms that may be used for the analysis of the network properties. Design of these algorithms that have polynomial execution time is difficult most of the time and heuristic algorithms or approximation algorithms that find suboptimal solutions in polynomial time are usually employed. Heuristic algorithms work well for most of the input combinations but do not guarantee a good solution for all inputs. Approximation algorithms on the other hand, guarantee a performance within a factor of the optimum value.

The main goal of this book is the identification, detailed description and application of a *repertoire* of algorithms that may be used for any complex network. Distance and centrality finding algorithms are one such class of algorithms that can be used for any complex network. Graph partitioning and clustering algorithms are another class of algorithms that detect communities and provide us with important information about the structures of them. Algorithms for detecting network motifs in general provide us with suboptimal solutions as this problem is NP-hard. Network alignment is another difficult problem encountered especially in complex biological networks.

1.5   Outline of the Book

The book consists of three parts. The first part is an overview of the background on graph theory and algorithms in Chapters 2 and 3 and the analysis of complex networks is investigated in Chapter 4.

Part II forms the core of the book with a distinct class of algorithms that may be used for any kind of complex network in each chapter. We first describe algorithms for distance finding and centrality in Chapter 5. Special subgraphs such as vertex cover, dominating sets and independent sets play an important role and can be used as building blocks for other complex network algorithms such as clustering; and are described in Chapter 6. Data clustering is the process of partitioning a data set into clusters and this topic is investigated in Chapter 7. Graph partitioning and graph clustering algorithms have important applications in all complex networks and can be used for community detection, as described in Chapter 8. Motif discovery algorithms and tools, algorithms and tools for finding frequently appearing motifs in complex networks are described in Chapter 9 along with the closely related subgraph isomorphism algorithms which find whether a graph contains the isomorphism of another graph as its subgraph.

Our focus in Part III is on implementation and we describe applications for each type of real-world complex network. We first start with biological networks and investigate clustering, motif finding and network alignment algorithms for protein-protein-interactive (PPI) networks in Chapter 10. We then describe social networks in detail and the implementation of community detection algorithms in these networks in Chapter 11. The technological networks using the Internet and the information networks using the Web examples are investigated in Chapter 12. Finally, we describe algorithms for clustering in wireless sensor networks along with an introduction to mobile social networks in Chapter 13.

A detailed study of complex networks is provided in [4]. The edited books by Dehmer [4] and Caldarelli and Vespignani [28] also have detailed analyses of complex networks. Analysis of biological networks is provided in [7] and a graph-theoretic view of complex networks is presented in [36].

References

[1] M. Dehmer (Ed.). *Structural Analysis of Complex Networks.* Birkenhauser, ISBN-13: 978-0-8176-4788-9, 2011.

[2] B.H. Junker and F. Schreiber (Eds.). *Analysis of Biological Networks.* Wiley Interscience, 2008.

[3] S. Milgram. The small world problem, *Psychology Today,* 1(1):61-67, 1967.

[4] M.E.J. Newman. *Networks, An Introduction.* Oxford University Press, ISBN-13: 978-0-1992-0665-0, 2010.

[5] G. Caldarelli and A. Vespignani (Eds.). *Large Scale Structure and Dynamics of Complex Networks.* World Scientific Publishing Co. Pte. Ltd., ISBN-13: 978-981-270-664-5, 2007.

[6] M.V. Steen. *Graph Theory and Complex Networks, An Introduction*, 2010.

BACKGROUND I

Chapter 2

Graph Theory

2.1 Basics

Graphs are discrete structures that connect points called *vertices* using lines called *edges*. Graphs are widely used to model networks where vertices represent the nodes of the network and the edges represent the interaction between the nodes. A graph $G(V,E)$ consists of a vertex set V, an edge set E and a relation that associates each edge with two vertices which are called the *endpoints* of the edge. A *loop* is an edge with the same endpoints and *multiple edges* are the edges with the same pair of endpoints. Figure 2.1 displays a graph where $V = \{a,b,c,d\}$ and $E = \{(a,b),(b),(b,c),(c,d),(a,c),(a,d),(a,c)\}$. Edge (b) is a loop and edges between vertices d and c are multiple edges. A *simple graph* does not have any loops or multiple edges.

The number of vertices of a graph $(|V|)$ is called its *order* and the number of its edges $(|E|)$ is called its *size*. In the context of this book, we will use the literals n for the order and m for the size of a graph. The complement of a graph $G(V,E)$ is $\overline{G}(V,E')$ with the same vertex set as G and $(u,v) \in E'$ if and only if $(u,v) \notin E$. A *clique* in a graph is a set of vertices such that each vertex in this set is connected to all other vertices in the set. Figure 2.2 shows the complement of a graph and a clique of order 5 (K_5).

Two vertices of a graph G are *neighbors* if there is an edge connecting these two vertices. The neighborhood of a vertex in a graph can be defined formally as follows.

$$N(v) = \{u \in V : \forall e(u,v) \in E\} \tag{2.1}$$

$N(v)$ is called the *open neighborhood* of v and $N[v] = N(v) \cup \{v\}$ is called the *closed neighborhood* of v, which is the union of all neighbors of v and itself. The closed neighborhood of vertex a in Figure 2.2.b is $\{a,c,d\}$.

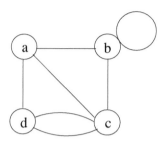

Figure 2.1: An example graph

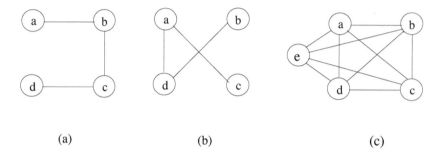

| (a) | (b) | (c) |

Figure 2.2: a) A graph with four vertices. b) Its complement. c) A clique of order 5

The *degree* of a vertex is the number of edges that have endpoints in that vertex. The maximum degree of a graph G is denoted by $\Delta(G)$ and the minimum degree of G by $\delta(G)$. $\Delta(G)$ of the graph in Figure 2.1 is 4 at vertex b (or c) as loops count twice and $\delta(G)$ is 3 for this graph. The average degree of a graph $G(V,E)$ is defined as follows:

$$deg_{av}(G) = \frac{1}{n} \sum_{v \in V} deg(v) \qquad (2.2)$$

The relation between the minimum, the average and the maximum degree of a graph can be stated as follows:

$$\delta(G) \leq deg_{av}(G) \leq \Delta(G) \qquad (2.3)$$

2.2 Subgraphs

Given two graphs $G(V,E)$ and $G'(V',E')$, if $V' \subseteq V$ and $E' \subseteq E$, G' is called a *subgraph* of G and G is called a *supergraph* of G'. If $\forall (x,y) \in E'$; $x \in V'$ and $y \in V'$,

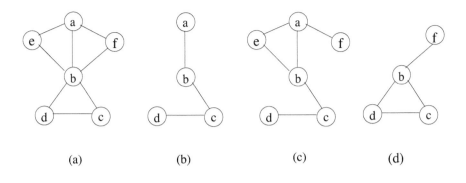

Figure 2.3: a) A graph. b) Its subgraph which is not induced. c) Its spanning subgraph. d) Its induced subgraph

then G' is an induced subgraph of G. An induced subgraph of a graph $G(V,E)$ can be obtained by deleting a vertex set $V_1 \in V$ along with any edges incident to vertices in this set. $G'(V',E') \subseteq G(V,E)$ is a *spanning subgraph* of G if $V' = V$. Figure 2.3 shows a graph, its spanning, induced and not induced subgraphs. A clique of a graph is also its induced subgraph.

The *union* of two graphs $G(V,E)$ and $G'(V',E')$ is the graph $H(V \cup V', E \cup E')$, that is, it has the union of the vertices and the union of the edges of the two graphs and it is shown as $H = G \cup G'$. The *intersection* of two such graphs is $S(V \cap V', E \cap E')$ which has the intersection of the vertices of the two graphs as its vertices and the intersection of the edges of the two graphs as its edges. Figure 2.4 displays the union and intersection of two graphs.

2.3 Graph Isomorphism

An isomorphism from a graph $G(V_1,E_1)$ to $H(V_2,E_2)$ is a bijection $f : V_1 \to V_2$ such that edge $(u,v) \in E_1$ if and only if edge $(f(u),f(v)) \in E_2$. These two isomorphic graphs are shown as $G \simeq H$. The graphs in Figure 2.5 are isomorphic where $f : a \to a', ..., g \to g'$.

Given two isomorphic graphs $G(V_1,E_1)$ and $H(V_2,E_2)$, V_2 contains all renamed elements of V_2. Two isomorphic graphs have the same graph theoretical properties. In the *subgraph isomorphism* problem, we are given two graphs G_1 and G_2, G_1 being a smaller graph than G_2, and we are asked to test the existence of a subgraph of G_2 that is isomorphic to G_1. We may be required to count the occurrences of graphs that are isomorphic to G_1 in G_2 and finding high frequencies of isomorphic graphs to G_1 in a complex network such as a biological network represented by G_2, may indicate a fundamental function performed by the module represented by G_1. Determination of graph isomorphism is NP-hard and subgraph isomorphism is NP-complete [6].

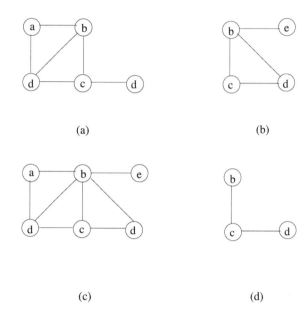

(a)   (b)

(c)   (d)

Figure 2.4: a) First graph. b) Second graph. c) Union of two graphs. d) Intersection of two graphs

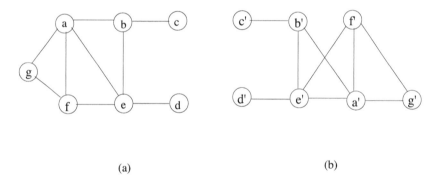

(a)   (b)

Figure 2.5: Two isomorphic graphs

2.4 Types of Graphs

Certain types of graphs have many applications in real complex networks. We will briefly describe bipartite, regular and weighted graphs as important graph types in this section.

A graph $G(V,E)$ is called *bipartite* if V can be partitioned into two subsets V_1 and V_2 such that each edge $(u,v) \in E$ has one endpoint in V_1 and the other endpoint in

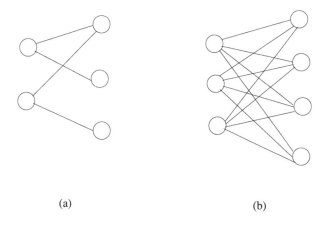

(a)   (b)

Figure 2.6: a) A 2-3 bipartite graph. b) A complete 3-4 bipartite graph

V_2. A bipartite graph G is *complete (k-l) bipartite* if given $|V_1| = k$, $|V_2| = l$; $\forall u \in V_1$ and $\forall v \in V_2$; $(u,v) \in E$. Figure 2.6 shows a 2-3 bipartite graph and a complete 3-4 bipartite graph.

A directed bipartite graph has directed edges between its two groups of vertices. The communities in the Web graph are usually represented by directed bipartite graphs and detection of these bipartite graphs provides the discovery of cyber communities in the Web.

A graph $G(V,E)$ is called *regular* if every vertex of G has the same degree. When degree is equal to k, then G is called k-regular. A *lattice* is a 4-regular graph except for the outer vertices which have a degree of 2 and a 3D hypercube is a 3-regular graph.

Edges are labeled with weights in a *weighted graph* $G(V,E,w)$ where $w : E \rightarrow \mathbb{R}$. These weights may be representing the importance of a link such as the importance of a relationship in a social network or the cost of sending a message between nodes in a communication network.

A *directed graph* or a *digraph* $G(V,E)$ consists of a nonempty set of vertices V and a set of directed edges E where each $e \in E$ is associated with an ordered set of vertices and an edge (u,v) is said to start from vertex u and end at vertex v [2]. It is clear that $(u,v) \in E$ of a digraph $G(V,E)$ does not imply $(v,u) \in E$. Figure 2.7 displays a weighted graph and a directed graph.

2.5 Paths and Cycles

A *walk* $w = (v_1, e_1, v_2, e_2, ..., v_n, e_n, v_n + 1)$ in a graph G is an alternating sequence of vertices and edges in V and E respectively such that for $1 \le i \le n$, $\{v_i, v_{i+1}\} = e_i$. A *path* of a graph $G(V,E)$ is its subgraph $G(V',E')$ where $V' = \{x_0, x_1, .., x_k\}$ and

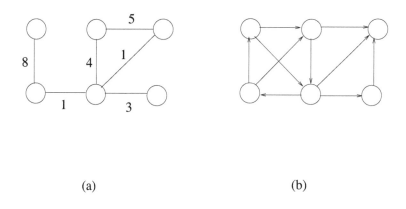

(a)   (b)

Figure 2.7: a) A weighted graph. b) A directed graph

$E' = \{(x_0,x_1),(x_1,x_2),...,(x_{k-1},x_k)\}$. A path is usually refereed by the sequence of its vertices. A *cycle* is a path that starts and ends at the same vertex and visits each vertex exactly once and has at least a length of 3. The length of a cycle is the number of vertices it has. The minimum length of a cycle in a graph G is called its *girth*, if there is a cycle. When G does not have any cycle, the girth is defined as 0. A *chord* of a graph $G(V,E)$ is an edge $e \in E$ which joins two vertices of a cycle in G but is not included in the cycle. A *Hamiltonian cycle* in a graph G is a cycle which includes all vertices of G. A *Hamiltonian path* of G is a path that visits each vertex of G exactly once. An *Eulerian path* in G is a path that contains each edge in G exactly once and an *Eulerian cycle* is a cycle that visits each edge in G exactly once.

The *distance $d(u,v)$* between the vertices u and v of a graph G is the length of the shortest path between u and v. The *diameter*, $diam(G)$, of a graph G is the longest distance between any two vertices of G. The *radius*, $rad(G)$, of G is defined as the smallest of the maximum distances between any two vertices.

The *eccentricity* of a vertex u of G is the maximum distance from u to any other vertex v. The radius is also the minimum eccentricity of the graph. For a graph $G(V,E)$, the *circumference* of a graph G is the length of the longest cycle, if there is a cycle in G. When G does not have any cycles, the circumference is defined as ∞. Figure 2.8 displays paths, a cycle and a chord in an example graph. The diameter of this graph is 3 as the shortest distance between two farthest nodes.

2.6 Connectivity

A graph G is *connected* if there exists a path between every pair of vertices, otherwise it is *disconnected*. The maximal connected induced subgraphs of G are called its *connected components* or simply *components*. A digraph G is *strongly connected* if for every walk from every vertex $v_1 \in V$ to any vertex $v_2 \in V$, there is also a walk

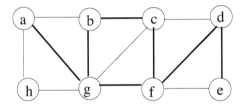

Figure 2.8: Paths and cycles in a graph where $\{a,g,f,d,e\}$ **is a path;** $\{b,c,f,g,b\}$ **is a cycle; edge** (c,g) **is a chord. The** $\{a,b,c,d,e,f,g,h,a\}$ **cycle is Hamiltonian, and** $\{a,h,g,b,c,f,e,d\}$ **is an Hamiltonian path**

from v_2 to v_1. Given a graph $G(V,E)$, an edge $e \in E$ is called a *bridge* if $\{G-e\}$ has more components than G. A minimal set of edges deletion of which disconnects G is called a *cutset* in G. Figure 2.9 displays a graph with two components and a cutset. A *cutpoint* of G is a vertex such that $\{G \setminus v\}$ has more components than G. A *block* of a graph G is its maximal connected subgraph which contains no cutpoints. Figure 2.10.a shows a graph with a bridge and a cutpoint $\{c\}$ is displayed in (b) removal of which results in three components as $\{a,b,h,g\}$, $\{d\}$ and $\{f,e\}$.

Vertex connectivity or just *connectivity* κ_v of a connected graph G is the minimum number of vertices that must be removed from G to have it disconnected or have only one vertex. In other words, it is the size of the minimum *vertex cut* it has. The graph of a computer network has vertices as the computational nodes and edges as the communication links between these vertices. A fundamental task in any computer network is the reliable transfer of messages among the nodes of the network and an obvious requirement from such a network to perform this task is that it should be connected. For this reason, we may be interested to know the number of minimum edge removals that will result in a disconnected network. *Edge connectivity* κ_e of a graph is exactly this parameter and if this is high, the network will not be disconnected until such number of links fail. Failure of many links simultaneously is

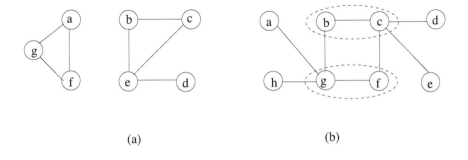

(a)   (b)

Figure 2.9: a) A graph with two components. b) A cutset of a graph $\{(b,c),(f,g)\}$

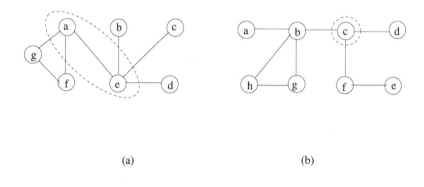

(a)   (b)

Figure 2.10: a) A bridge ({ae}) of a graph. b) A cutpoint ({c}) of a graph

less probable and we can therefore say that a computer network with a higher connectivity is more reliable than a network with lower connectivity. The edge connectivity of a graph with a bridge is 1.

2.7   Trees

A graph without a cycle is called *acyclic*. An acyclic graph with more than one component is called a *forest* and a *tree* is a connected acyclic graph. Directed trees and forests are acyclic directed graphs. Trees have many applications in computer science and bioinformatics. The following are equivalent to describe a tree T [2]:

■ T is a tree;

■ T contains no cycles, and has $n - 1$ edges;

■ T is connected, and has $n - 1$ edges;

■ T is connected, and each edge is a bridge;

■ Any two vertices of T are connected by a exactly one path.

A spanning tree T of a graph $G(V, E)$ covers (spans) all vertices of G. In *rooted trees*, there is a special vertex called the *root* and the vertices have an orientation towards the root. The *parent* of a vertex u in such a tree is the vertex that u is connected on the path to the root. A vertex u is a *child* of a vertex v if v is its parent. A *leaf* vertex does not have any child vertices and the root vertex does not have a parent.

Given a weighted graph $G(V, E, w)$, a spanning tree T of G is called a *minimum spanning tree (MST)* of G, if total sum of the weights of its edges is minimum among all possible spanning trees of G. If all weights of the edges of a graph G are unique, then there is exactly one spanning tree of G. Figure 2.11.a displays a spanning tree of a graph without any root and a rooted minimum spanning tree is shown in (b) with the root vertex e where each child node points to its parent.

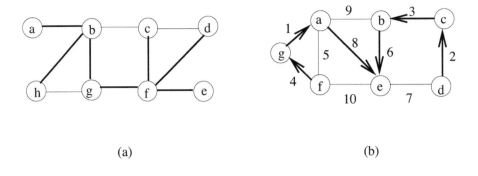

(a) (b)

Figure 2.11: a) A spanning tree. b) A minimum spanning tree rooted at vertex *e*

2.8 Graph Representations

Although plain figures are easy to visually inspect the properties of small graphs, we need to represent larger graphs in other forms to be able to perform some computation on them. Two widely used methods of representation are the *adjacency matrices* and *adjacency lists*. The adjacency matrix of a graph $G(V,E)$ with n vertices is an $n \times n$ matrix which has elements a_{ij} defined as follows:

$$a_{ij} = \begin{cases} 0 & \text{if } i = j \\ 1 & \text{if vertex } v_i \text{ is connected to vertex } v_j \end{cases} \qquad (2.4)$$

The adjacency list of a graph $G(V,E)$ with n vertices on the other hand, is a list of n elements where each element consists of a vertex $v \in V$ and its neighbors in a linked list. The *incidence matrix* of a graph $G(V,E)$ with n vertices and m edges is an $n \times m$ matrix has elements a_{ij} defined as follows.

$$a_{ij} = \begin{cases} 0 & \text{if vertex } v_i \text{ is not incident to edge } (v_i, v_j) \\ 1 & \text{if vertex } v_i \text{ is incident to edge } (v_i, v_j) \end{cases} \qquad (2.5)$$

Furthermore, the $n \times n$ distance matrix D of a weighted graph $G(V,E,w)$ with n vertices can be defined where each entry d_{ij} of D is equal to the distance between the nodes v_i and v_j. Figure 2.12 displays the adjacency matrix, the adjacency list and the incidence matrix of a graph.

2.9 Spectral Properties of Graphs

Spectral properties of a graph involve finding the eigenvalues and eigenvectors of matrices associated with the graph from which its connectivity can be investigated and assessed.

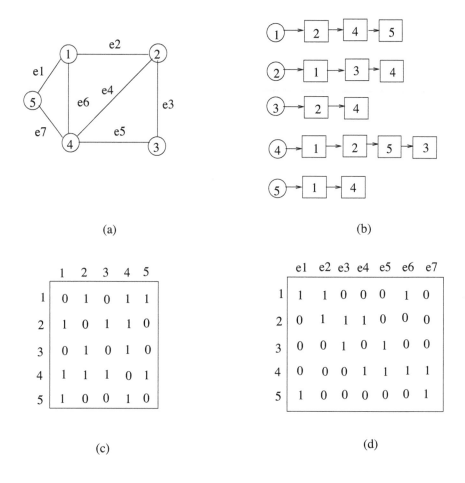

(a)

(b)

(c)

(d)

Figure 2.12: a) An example graph. b) Its adjacency list. c) Its adjacency matrix. d) Its incidence matrix

2.9.1   Eigenvalues and Eigenvectors

An *eigenvalue* λ and an *eigenvector* x of a square matrix A satisfy the equation

$$Ax = \lambda x \tag{2.6}$$

We call the nonzero vector x as the eigenvector corresponding to eigenvalue λ. Eqn. (2.6) can be rewritten as,

$$Ax - \lambda x = 0 \tag{2.7}$$

$$(A - \lambda I)x = 0 \tag{2.8}$$

Therefore, x is an eigenvector of A if and only if $\det(A - \lambda I) = 0$. This polynomial is called the *characteristic equation* of matrix A. Finding eigenvalues of square matrices has many applications in engineering, physics, sociology and statistics. In order to find eigenvalues and eigenvectors of a matrix, we can proceed as follows:

1. Form matrix $(A - \lambda I)$.

2. Solve the characteristic equation for λ values.

3. Substitute each eigenvalue in Eqn. (2.6) to find x vectors corresponding to λ values.

We will show an example of this method for he matrix:

$$A = \begin{pmatrix} 7 & 2 \\ -10 & -2 \end{pmatrix} \tag{2.9}$$

$(A - \lambda I)$ becomes:

$$\begin{pmatrix} 7 - \lambda & 2 \\ -10 & -2 - \lambda \end{pmatrix} \tag{2.10}$$

The characteristic equation of A is $\lambda^2 - 5\lambda + 6$. Solving for λ yields $\lambda_1 = 2$ and $\lambda_2 = 3$. Substituting λ_1 in Eqn. (2.7) results in:

$$\begin{pmatrix} 5 & 2 \\ -10 & -4 \end{pmatrix} \begin{pmatrix} x_1 \\ x_2 \end{pmatrix} = 0 \tag{2.11}$$

and we find $X_1 = [-0.41]$. Similarly, for λ_2, the eigenvector X_2 is $[-0.4\ 1]$. Finally, we can test whether the solutions are valid. We will do this for λ_1 and X_1 by substituting these values in (2.6) as follows:

$$\begin{pmatrix} 7 & 2 \\ -10 & -2 \end{pmatrix} \begin{pmatrix} -0.4 \\ 1 \end{pmatrix} = 2 \begin{pmatrix} -0.4 \\ 1 \end{pmatrix} \tag{2.12}$$

which is consistent with the values found.

2.9.2 The Laplacian Matrix

The Laplacian matrix of a graph represents its connectivity and is sometimes called the *connectivity matrix* of the graph. The Laplacian matrix L is equal to $D - A$ where D is the degree matrix and A is the adjacency matrix. An entry of the Laplacian therefore is given by:

$$L_{ij} = \begin{cases} deg(i) & \text{for} \quad i = j \\ -a_{ij} & \text{for} \quad i \neq j \end{cases} \tag{2.13}$$

The Laplacian matrix of the graph of Figure 2.12.a is as follows:

$$\begin{pmatrix} 3 & -1 & 0 & -1 & -1 \\ -1 & 3 & -1 & -1 & 0 \\ 0 & -1 & 2 & -1 & 0 \\ -1 & -1 & -1 & 4 & -1 \\ -1 & 0 & 0 & -1 & 2 \end{pmatrix}$$

For an undirected graph G, The normalized Laplacian $\mathcal{L}$ is defined as follows:

$$L_{ij} = \begin{cases} 1 & \text{if } i = j \text{ and } deg(j) \neq 0 \\ -\dfrac{1}{\sqrt{deg(i)deg(j)}} & \text{if } i \text{ and } j \text{ are adjacent} \\ 0 & \text{otherwise} \end{cases} \qquad (2.14)$$

which can be written as:

$$\mathcal{L} = D^{-1/2}(D - A)D^{-1/2} \qquad (2.15)$$

Given a graph G and its Laplacian matrix L with eigenvalues $\lambda_0 \leq \lambda_1, ..., \lambda_{n-1}$, first thing to note is that L is always positive-semidefinite meaning $\forall i, \lambda_i \geq 0$ and $\lambda_0 = 0$. The eigenvalues of the L display important properties about connectivity of graph G. The number of 0 values gives the number of connected components of G. The second smallest eigenvalue of L is called the *algebraic connectivity* of G and is named the *Fiedler value* of G [9]. This eigenvalue is greater than 0 if and only if G is connected. The magnitude of algebraic connectivity is related to how well G is connected. For a graph with n vertices, the algebraic connectivity has an upper bound of $1/(n.diam(G))$ [8].

2.10   Chapter Notes

In this chapter, we have reviewed basic concepts in graph theory in regards to complex network applications. Finding clusters in complex networks is a fundamental problem and various algorithms for this task are employed. Cliques or clique-like structures are also searched in complex networks as these may indicate the existences of dense regions of complex networks. Detecting specific subgraph structures that occur more than usual is a motif finding problem and a fundamental research topic especially in biological networks. In this case, we will be searching for subgraph isomorphism where subgraphs of a target graph that are isomorphic to each other and sometimes to a given small input graph are searched. Spectra of graphs and the Laplacian matrix play an important role in finding centralities and detecting clusters as well as partitioning graphs as we shall see in Part II.

In-depth review of graph theory can be found in the classical book on graph theory by Harary [4] and a more updated view is in the book by West [9]. Two more recent books are by Bondy and Murty [1] and Fournier [2]. We will investigate more graph properties related to the degrees, clustering and distances of vertices in graphs in Chapters 4 and 5.

Exercises

1. Show the components, cliques and blocks of size greater than 2 for the graph of Figure 2.13.

2. For the graph of Figure 2.14, investigate the existences of Eulerian paths, Eulerian cycle, Hamiltonian paths and Hamiltonian cycle.

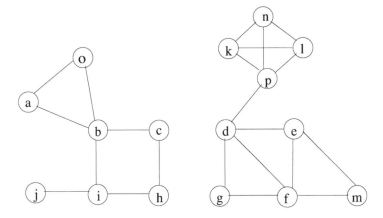

Figure 2.13: The example graph for Ex. 1

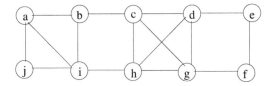

Figure 2.14: The example graph for Ex. 2

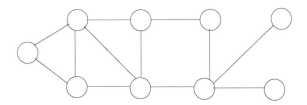

Figure 2.15: The example graph for Ex. 3

3. For the graph of Figure 2.15 work out the radius, diameter, girth and circumference.

4. Show all the spanning trees of the graph in Figure 2.16. Which one is the MST?

5. Find the eigenvalues and the eigenvectors of the following matrix:

$$L = \begin{pmatrix} 4 & 1 \\ 3 & 6 \end{pmatrix}$$

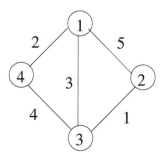

Figure 2.16: The example graph for Ex. 4 and 10

6. For a bipartite graph $G(V_1, V_2, E)$ where V_1 and V_2 are the disjoint vertex sets, show that $\sum_{u \in V_1} deg(u) = \sum_{v \in V_2} deg(v)$.

7. Show that for any graph G, $rad(G) \leq diam(G) \leq 2rad(G)$.

8. Find the radius, girth, circumference and the diameter of the complete bipartite graph $K_{m,n}$ in terms of m and n.

9. Show that every tree with n nodes has $n - 1$ edges.

10. Work out the Laplacian matrix and the algebraic connectivity for the graph of Figure 2.16.

References

[1] J.A. Bondy and U.S.R. Murty. *Graph Theory*. Springer Graduate Texts in Mathematics, ISBN 978-1-84628-970-5, 2008.

[2] J.C. Fournier. *Graph Theory and Applications*. Wiley, ISBN 978-1-848321-070-7, 2009.

[3] K. Erciyes. *Distributed Graph Algorithms for Computer Networks*. Computer Communications and Networks Series, Springer, ISBN 978-1-4471-5172-2, 2013.

[4] M. Fiedler. Laplacian of graphs and algebraic connectivity. *Combinatorics and Graph Theory*, 25:57-70, 1989.

[5] M.R. Garey and D.S. Johnson. *Computers and Intractability: A Guide to the Theory of NP-Completeness*. W. H. Freeman, 1979.

[6] C. Griffin. *Graph Theory*. Penn State Math 485 Lecture Notes. Homepage: http://www.personal.psu.edu/cxg286/Math485.pdf, 2011.

[7] F. Harary. *Graph Theory*. Addison-Wesley, 1979.

[8] B. Mohar. The Laplacian Spectrum of Graphs. In *Graph Theory, Combinatorics, and Applications*, Ed. Y. Alavi, G. Chartrand, O. R. Oellermann, A. J. Schwenk, Wiley, 1991, 2:871-898, 1991.

[9] D.B. West. *Introduction to Graph Theory*, 2nd Ed., Prentice-Hall, ISBN 0-13-014400-2, 2001.

Chapter 3

Algorithms and Complexity

3.1 Introduction

An algorithm is a sequence of instructions that provides a solution to a given problem. In general, an algorithm works on an *input* and provides an *output* by applying the instructions on this input. Clearly, a fundamental requirement for any algorithm is that it should work correctly. Also, we would be interested to know the time needed to solve the problem, which is expressed as the number of steps for the algorithm to terminate. Another point of interest for the algorithm designer and the implementor is whether any other algorithm with better performance, that is, with fewer steps exists.

An *expression* in an algorithm is built using constants and variables. *Statements* are composed of expressions and are the main units of executions. Statements can be mainly of three types; an *assignment*, a *control* or a *repetition* statement. An assignment assigns a value to a variable as follows:

$$\textbf{int} \quad a \leftarrow 12$$

Here, we declare a variable of type integer and initialize it with the value 12. In order to branch within an algorithm, the **if** *condition*, **then** *statement* 1, **else** *statement* 2 structure can be used, where *condition* is a Boolean expression which either evaluates to *true* or *false*. If the result of the evaluation is *true*, *statement* 1 is executed, otherwise execution resumes with *statement* 2.

We frequently need repetition in an algorithm where the same operation is performed many times on different data. The *for*, *while* and *loop* constructs provide the necessary structures for repetition. The *for* loop is used when the count of iterations is known before entering the loop as follows, where an integer a is incremented n times:

1. **for** $i \leftarrow 1$ **to** n **do**

2. $a \leftarrow a + 1$

3. **end for**

The **for all** loop has a similar construct but selects its elements arbitrarily from a set and works until all members of the set are processed as shown below, where a set P with three integer elements is given and each element of P is replaced by its consecutive integer value.

1. $P \leftarrow \{5, 2, 9\}$

2. **for all** $u \in P$ **do**

3. $P \leftarrow P \setminus \{u\}$

4. $P \leftarrow P \cup \{u + 1\}$

5. **end for**

For the cases where the count of the execution of the loop is not known beforehand, the *while-do* construct may be used in which the Boolean expression is evaluated and the loop is entered if this value is *true* as shown below where the loop runs until the value 5 is entered.

1. **input** a

2. **while** $a \neq 5$ **do**

3.   **input** a

4. **end while**

We may need to execute a loop at least once and *loop-until* construct may be used for this purpose where the test condition is is checked at the end of the loop. Re-writing the previous example with this construct results in one fewer statement as follows:

1. **loop**

2.   **input** a

3. **until** $a \neq 5$

In this chapter, we will investigate fundamental algorithm design methods and for the problems where no solution with favorable performance is known, we will look at ways of designing algorithms that approximate a solution to the problem. We conclude by the introduction of parallel and distributed algorithms where more than one computation element is used in solving a problem.

3.2 Time Complexity

The time complexity of an algorithm is the number of steps required until the desired output is obtained. Let us demonstrate this concept by using an example algorithm where our aim is to find the maximum value of numbers in an array *a*. The general idea behind this algorithm is to have a variable called *max* which will contain the value of the maximum element and another variable called *index* which will show its index. We start by equating *max* to the first element of *a* and then compare the value in *max* with all elements of *a*. If the compared array value is greater than *max*, that value is copied to *max* and its index *i* stored in *index* as shown in Alg. 3.1.

Algorithm 3.1 *Find_Max*

1: **real** $a[8] = \{7,1,6,3,2,4,8,5\}$ ▷ integer array *a* of 8 elements is declared
2: **int** *max*, *index* ▷ output variables are declared
3: $max \leftarrow a[1], index \leftarrow 1$ ▷ initialize output variables
4: **for** $i \leftarrow 2$ to n **do** ▷ compare *max* against all *a* values
5:   **if** $a[i] > max$ **then**
6:     $max \leftarrow a[i]$
7:     $index \leftarrow i$
8:   **end if**
9: **end for**
10: **return** *max*

The output variables *max* and *index* will have values 8 and 7 consecutively upon termination for the given input values and it can be easily seen that this algorithm requires $n-1$ steps for the execution of the *for* loop.

In general, the computation time of an algorithm is expressed as the number of steps as a function of its input size. However, precision is not needed in most cases. For example, if we compute the time complexity of an algorithm as $4n^3+12n^2+23$ steps, it is sensible to discard all other terms than the highest exponential one for large *n* values, resulting in $4n^3$ as this term will have a dominant effect on the number of steps. Furthermore, the constants can also be discarded and we say the time complexity of this algorithms is $O(n^3)$ which is to say that the running time grows as fast as n^3 and is much simpler than the original expression of complexity. Let us assume f and g are two functions from $\mathbb{N}$ to $\mathbb{R}^+$.

■ $f(n) = O(g(n))$, if there is a constant $c > 0$ such that $f(n) \le cg(n)$ $\forall n \ge n_0$.

■ $f(n) = \Omega(g(n))$, if there is a constant $c > 0$ such that $f(n) \ge cg(n)$ $\forall n \ge n_0$.

■ $f(n) = \Theta(g(n))$, if $f(n) = O(g(n))$ and $f(n) = \Omega(g(n))$.

The $O(g(n))$ states that when the input size is equal to or greater than a threshold n_0, the running time is always less than $cg(n)$. In other words, the $g(n)$ function

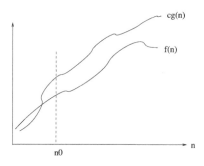

Figure 3.1: $O(g(n))$ **function**

provides an upper bound on the running time. As an example, we claim $20n + 4 \in O(n^2)$. $20n + 4 \le 20n + n$, $\forall n \ge 4$. We can therefore state that $21n \le 21n^2$ and using transitivity, $20n + 4 \le 21n^2$, $\forall n \ge 4$, which yields 21 and 4 as c and n_0. An example $O(g(n))$ is shown in Figure 3.1.

On the contrary, the $\Omega(g(n))$ provides a lower bound on the running time, ensuring that the algorithm requires at least $cg(n)$ steps after a threshold n_0 as shown in Figure 3.2. As an example, we claim $n^3 \in \Omega(n^2)$. Since $n^3 \ge n^2$, $\forall n \ge 0$ yielding $c = 1$ and $n_0 = 0$.

If there are two constants $c_1, c_2 > 0$, such that the running time of the algorithm is greater than or equal to $c_1.g(n)$ and less than or equal to $c_2.g(n)$ after a threshold n_0; the running time of the algorithms is $\Theta(g(n))$. For example, we claim $n^2 + 3n \in \Theta(n^2)$; $n^2 + 3n \ge n^2$, $\forall n \ge 1$ and $n^2 + 3n \le (n^2 + 3n^2 = 4n^2)$ $\forall n \ge 1$. Therefore, $g(n) = n^2$, c_1 is 1, c_2 is 4 and n_0 is 1. An example $\Theta(g(n))$ is shown in Figure 3.3. In other words, the running time of the algorithm is bounded by $c_1.g(n)$ and $c_2 g(n)$.

Some general rules for the upper bound on time complexity are as follows:

■ Constant factors can be discarded. For example, $8\ n^2 \to n^2$.

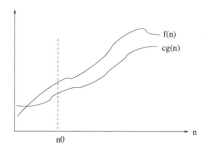

Figure 3.2: $\Omega(g(n))$ **function**

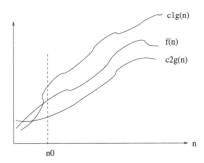

Figure 3.3: $\Theta(g(n))$ **function**

- n^a dominates n^b if $a > b$. For example, $n^5 \to n^3$.

- Any exponential dominates any polynomial. For example, $2^n \to n^5$.

- Any polynomial dominates any logarithm. For example, $n^2 \to n \log n$.

3.3 Recurrences

An algorithm that calls itself for a certain number of times is called *recursive*. Recursive algorithms require a *base case* to start returning to the calling instant of the algorithm. These algorithms are simpler but their analysis, in general, is more difficult than analysis of iterative algorithms. We will describe a simple recursive algorithm that computes the factorial of an integer. The factorial $F(n)$ of a positive integer n is defined as the product of all integers $1,...,n$ and $F(n) = F(n-1) \times n$ for any value of n. Alg. 3.2 shows how to implement this recursive algorithm.

Algorithm 3.2 *Fact_rec*

1: **Input** : a positive integer n
2: **Output** : $n!$
3: **if** $n = 0$ **then**
4:      **return** 1
5: **else   return** $n \times F(n-1)$
6: **end if**

If $T(n)$ is the number of multiplications,

$$T(n) = T(n-1) + 1 \qquad (3.1)$$

Such equations where a function is defined in terms of its value for another input are called *recurrence relations* or *recurrences*. There may be more than one solution to recurrences as in Eqn. (3.1), but when the initial condition $T(0) = 0$ is considered, we will find one solution. We can use a method called *backward substitution* to solve Eqn. (3.1) where values of n are replaced by its previous values as follows:

$$T(n) = T(n-1) + 1 \tag{3.2}$$
$$= [T(n-2) + 1] + 1 = T(n-2) + 2$$
$$= [T(n-3) + 1] + 2 = T(n-3) + 3$$

As can be seen, there is an emerging pattern. We would still need mathematical proofs that this pattern is valid for any n. Substituting the initial condition in this pattern yields:

$$T(n) = T(n-1) + 1 = T(n-i) + i \tag{3.3}$$
$$= T(n-n) + n = T(0) + n = n$$

This may have been easily noticed prior to the analysis, however, we would still need to show that this is indeed the case.

3.4 Divide and Conquer Algorithms

The divide and conquer method of designing algorithms first divides the problem into a number of subproblems which are then usually solved recursively in the second step. The final step consists of combining the answers obtained in step 2 to get the actual output. Typically in a divide and conquer strategy, a problem of size n is divided into instances of size n/b with a instances to be solved where $a \geq 1, b \geq 1$. For simplicity, assuming n is a power of b results in the following recurrence relation [7]:

$$T(n) = aT(n/b) + f(n) \tag{3.4}$$

where $f(n)$ shows the time for dividing the problem into subproblems and merging the solutions. Given such a recurrence which in fact describes many divide and conquer algorithms, the following theorem is used to calculate the time complexity of the algorithm.

Theorem 1 (Master Theorem) *If $f(n) \in \Theta(n^d)$ with $d \geq 0$ in a recurrence equation of the form as in Eqn. (3.4), then,*

$$T(n) = \begin{cases} \Theta(n^d) & \text{if} \quad a \leq b^d \\ \Theta(n^d \log n) & \text{if} \quad a = b^d \\ \Theta(n^{\log_b^a}) & \text{if} \quad a \geq b^d \end{cases} \tag{3.5}$$

We will illustrate the use of this method by implementing the *mergesort* algorithm. The idea of the algorithm is to recursively divide the array of elements into two halves; sorting the elements in the halves recursively and merging the two sorted subarrays into a single sorted one. The recursive function *mergesort* is shown in Alg. 3.3 as in [7] and Figure 3.4 displays the execution of the *mergesort* algorithm on an array of size 8.

Algorithm 3.3 *Mergesort*

1: **procedure** *mergesort*$((a[n])$
2:     **Input** : $a[n]$ ▷ array $a[0,...,n-1]$
3:     **Output** : sorted $a[n]$ in non-decreasing order
4:     **if** $n > 1$ **then**
5:         $b[0,...,\lfloor n/2 \rfloor - 1] \leftarrow a[0,...,\lfloor n/2 \rfloor - 1]$
6:         $c[0,...,\lfloor n/2 \rfloor - 1] \leftarrow a[\lfloor n/2 \rfloor,...,n-1]$
7:         $mergesort(b[0,...,\lfloor n/2 \rfloor - 1)$
8:         $mergesort(c[0,...,\lfloor n/2 \rfloor - 1)$
9:         $merge(b,c,a)$
10:    **end if**
11: **end procedure**
12: **procedure** *merge*$((b,c,a)$
13:     **Input** : sorted arrays **int** $b[0,...,\lfloor l \rfloor - 1], c[1..\lfloor m \rfloor - 1]$
14:     **Output** : sorted array $a[0..l+m-1]$
15:     **int** $i,j,k \leftarrow 0$
16:     **while** $i \leq l \wedge j \leq m$ **do**
17:         **if** $b[i] \leq c[j]$ **then**
18:             $a[k] \leftarrow b[i]; i \leftarrow i+1$
19:         **else**$a[k] \leftarrow c[j]; j \leftarrow j+1$
20:         **end if**
21:         $k \leftarrow k+1$
22:         **if** $i = l$ **then**
23:             $a[k,...,l+m-1] \leftarrow c[j,...,m-1]$
24:         **else**$a[k,...,l+m-1] \leftarrow b[i,...,l-1]$
25:         **end if**
26:     **end while**
27: **end procedure**

The *mergesort* algorithm has a recursive structure and the recurrence is given by:

$$T(n) = 2T(n/2) + T_{merge}(n) \text{ for } n \geq 1, T(1) = 0. \tag{3.6}$$

The worst case for T_{merge} is $n - 1$. Substituting this value and using the master theorem yields $T(n) = \Theta(n \log n)$ for the time complexity of *mergesort* algorithm which is similar to a general comparison based sorting algorithm. Divide and conquer method is suited for parallel computations where a problem is divided into many subproblems and each parallel processing element solves its own subproblem concurrently.

3.5   Graph Algorithms

Graph algorithms are used for problems which make use of the graph structure we have seen. We will use graphs frequently to represent complex networks. Traversals of graphs can be accomplished by visiting their vertices in some defined order. Two fundamental methods for graph traversal are the breadth-first search and depth-first search which are described next.

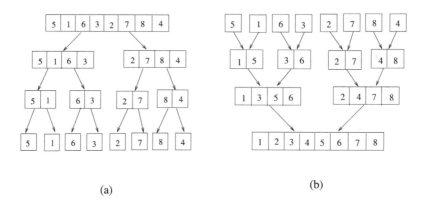

(a)   (b)

Figure 3.4: Mergesort algorithm example execution. a) Divide step. b) Merge step

3.5.1 Breadth-first Search

For unweighted graphs, distances from a source node s can be calculated by proceeding layer by layer. We can label all nodes that are direct neighbors of s with distance 1 and all nodes adjacent to these one-hop nodes can have the distance label of 2, continuing until all nodes of the graph G are labeled. *Breadth-first search* (BFS) implements this procedure by iteratively labeling nodes. Alg. 3.4 shows one way of

---

Algorithm 3.4 *BFS*

1: **Input** : $G(V,E)$
2: **Output** : d_v and $pred[v]$ $\forall v \in V$ ▷ shows distance and place of a vertex in BFS tree
3: **for all** $u \in V \setminus \{s\}$ **do**   ▷ initialize distances
4:      $d_u \leftarrow \infty$
5:      $pred[u] \leftarrow \perp$
6: **end for**
7: $d_s \leftarrow 0$
8: $pred[s] \leftarrow s$
9: $Q \leftarrow s$
10: **while** $Q \neq \emptyset$ **do**   ▷ continue until Q is empty
11:      $u \leftarrow deque(Q)$ ▷ get the first element
12:      **for all** $(u,v) \in E$ **do**   ▷ check all neighbors
13:          **if** $d_v = \infty$ **then**
14:              $d_v \leftarrow d_u + 1$
15:              $pred[v] \leftarrow u$
16:              $enque(Q,v)$
17:          **end if**
18:      **end for**
19: **end while**

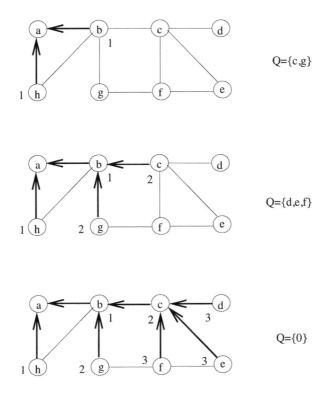

Figure 3.5: BFS algorithm execution

implementing the BFS algorithm where Q is a queue which contains the neighbors of currently labeled nodes and the array *pred*[*n*] shows the position of a node in the BFS tree by pointing to its parent and d_v is the distance of a node v to the source node s.

The algorithm starts from the source vertex s and places it in the queue Q. It then iteratively removes an element u from Q, and if neighbors of u are not yet labeled, labels them with distance that is one more than the distance of u to the source vertex s. Proceeding in this manner guarantees each neighbor vertex v of the vertex u is one step further from vertex s, hence providing the BFS property. Figure 3.5 displays an example network where BFS algorithm is executed and the labeled node at each iteration points to its parent along the shortest path to the source node.

Theorem 3.1
The time complexity of BFS algorithm is $\Theta(n+m)$ for a graph of order n and size m.

Proof 2 *The initialization between lines 3 and 6 takes $\Theta(n)$ time. The while loop is executed at most n times and the for loop between the lines 12 and 18 is run at most $deg(u)+1$ times considering the vertices with no neighbors. Total running time in this* *case is:*

$$n + \sum_{u \in V}(deg(u) + 1) = n + \sum_{u \in V} deg(u) + n = 2n + 2m \in \Theta(n + m)$$

3.5.2 Depth-first Search

Depth-first search (DFS) of a graph is an efficient method that provides important information such as connectivity about graphs. It can be used for both directed and undirected graphs. The general idea of this method is to start from a vertex and visit one of its neighbors and a neighbor of this neighbor by marking all visited nodes in process until no unmarked node is found.

Alg. 3.5 shows one way of implementing the DFS algorithm for a graph $G(V,E)$ that may have disconnected components. All of the nodes of G are first marked as unvisited and their predecessors which will show their parents in the DFS tree are undefined. Then for each unvisited node u of G, DFS search is started by calling the recursive DFS procedure in line 10 of the main algorithm. This procedure will traverse all of the nodes of the connected component containing u and starts by marking

---

Algorithm 3.5 *DFS_forest*

 1: **Input** : $G(V,E)$, directed or undirected
 2: **Output** : $pred[n]$; $firstvis[n]$, $secvis[n]$ ▷ shows place of a vertex in DFS tree
 3: **int** $time \leftarrow 0$; $visited[1{:}n] \leftarrow 0$
 4: **for all** $u \in V$ **do**   ▷ initialize distances
 5:      $visited[u] \leftarrow false$
 6:      $pred[u] \leftarrow \perp$
 7: **end for**
 8: **for all** $u \in V$ **do**
 9:      **if** $\neg visited[u]$ **then**
10:           $DFS(u)$ ▷ call for each connected component
11:      **end if**
12: **end for**
13:
14: **procedure** $DFS(u)$
15:      $visited[u] \leftarrow true$
16:      $time \leftarrow time + 1$; $firstvis[v] \leftarrow time$ ▷ first visit
17:      **for all** $(u,v) \in E$ **do**   ▷ visit neighbors
18:           **if** $\neg visited[v]$ **then**
19:                $pred[v] \leftarrow u$
20:                $DFS(v)$
21:           **end if**
22:      **end for**
23:      $time \leftarrow time + 1$; $secvis[v] \leftarrow time$ ▷ return visit
24: **end procedure**

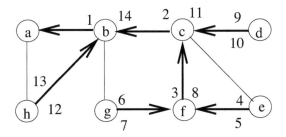

Figure 3.6: DFS algorithm execution

node *u* as visited and recording its visit time. An integer variable *time* is incremented each time a node *v* is visited whether first time or upon completion of the visits to all neighbors of *v*.

Figure 3.6 displays the execution of *DFS_forest* algorithm in the same example network Figure 3.5 where each node has the first and last visit times shown. There is a single connected component in this graph and therefore the loop between lines 8 and 11 is executed once and the total time is 16.

The initialization of the DFS algorithm takes $\Theta(n)$ time as each vertex is considered. The main recursive part has a time complexity of $\Theta(m)$ as every edge must be considered twice during the inspection of the neighbor vertices of every vertex. Total time taken therefore is $\Theta(n+m)$.

3.6   Dynamic Programming

Dynamic programming is an algorithmic method to solve problems that have overlapping subproblems. It is a powerful method used in algorithms for optimization problems. In this method, the subproblems are solved once and the solutions are saved to be used for the solution of the original problem. The basic steps of dynamic programming are as follows:

1. Decomposing the problem into smaller and simple subproblems.

2. Storing the solutions to the subproblems in a table to be reused.

3. Combining the solutions of the smaller problems to solve larger problems.

Decomposition of the problem into smaller problems is the main step of dynamic programming and is called the formulation of the problem. Problems to be solved by dynamic programming can usually be defined by recurrences. As an example, let us consider the Fibonacci numbers which are a sequence as follows:

$$0, 1, 2, 3, 5, 8, 13, 21, 34, 55$$

which can be defined by the recurrence:

$$F(n) = F(n-1) + F(n-2) \qquad for \quad n \geq 2 \qquad (3.7)$$

with the initial condition $F(0) = 0$ and $F(1) = 1$. In order to compute $F(n)$, we need to compute $F(n-1)$ and $F(n-2)$ which have overlapping subproblems. Therefore, instead of solving the recurrence, we can start by the initial values of 0 and 1; and store the values computed thereafter in an array to be used for further computations. Alg. 3.6 shows how to compute Fibonacci numbers using dynamic programming which takes $O(n)$ time.

Algorithm 3.6 *Fibo_dyn*

1: **Input** : $n \geq 2$
2: **Output** : nth Fibonacci number
3: **int** $F[n]$
4: $F[0] \leftarrow 0; F[1] \leftarrow 1$
5: **for** $i \leftarrow 2$ to n **do**
6:     $F[i] \leftarrow F[i-1] + F[i-2]$
7: **end for**
8: **return** $F(n)$

As another example, let us consider the *longest increasing subsequence problem* (LIS). Given a sequence $S = \{a_1, a_2, .., a_n\}$ of numbers, L is the longest subsequence of S where $\forall a_i, a_j \in L, a_i \leq a_j$ if $i \leq j$. For example, given $S = \{7, 2, 4, 1, 6, 8, 5, 9\}$, $LS = \{2, 4, 6, 8, 9\}$. A dynamic programming solution to the problem would again start from the subproblems and store the results to be used in future. Alg. 3.7 shows how to find LIS using dynamic programming in $O(n^2)$ time. We have used the algorithm that finds the maximum of numbers in an array (*Find_Max* of Section 3.2) to find the length of the longest path stored in the array L.

Algorithm 3.7 *LIS_dyn*

1: **Input** : $S[n]$ ▷ array of numbers
2: **Output** : length   ▷ length of LIS
3: **for** $i \leftarrow 1$ to n **do**
4:     $L[i] \leftarrow 1$
5:     **for** $j \leftarrow 1$ to i **do**
6:         **if** $S[i] \geq S[j] \wedge L[i] \leq L[j]$ **then**
7:             $L[i] \leftarrow L[j] + 1$
8:         **end if**
9:     **end for**
10: **end for**
11: **return** *Find_Max*(L)

In these two examples, we have seen how to implement dynamic programming using a *bottom-up approach* where we started from the subproblems and worked to find the solutions to the larger problems until we obtained the solution to the original problem. In the *top-down approach*, memory functions are used to solve unnecessary subproblems. Dynamic programming is different from the divide and conquer method as the latter finds solutions to non-overlapping problems. Dynamic programming is a powerful technique which is widely used for bioinformatic problems.

3.7 Greedy Algorithms

A *greedy* algorithm makes a locally optimal choice at each step with the expectation of finding a global optimum. A greedy method in general, may not provide an optimal solution to the problem at hand. However, it may approximate a global optimal solution in a reasonable time. As an illustrative example of greedy algorithms, we will implement an algorithm due to Prim called *Prim_MST* to find the MST of a weighted graph $G(V,E,w)$. We will first define few concepts and properties as follows [2].

Definition 3.1 Minimum Weight Outgoing Edge Let $T(V',E',w) \subseteq G(V,E,w)$ be an MST of graph G. A fragment $F \in$ MST is a subtree of the MST. The minimum weight outgoing edge (*MWOE*) of a given F is an edge $(u,v) \in E$ such that $u \in F$ and $v \notin F$ with the minimum weight among all outgoing edges from F.

We can now state two important properties of MSTs.

Lemma 3.1 Property 1: blue rule
Given a fragment of F_a of an MST, let (u,v) with $u \in F$ and $v \notin F$ be an MWOE of F_a. Then joining (u,v) to F_a yields another fragment F_b of the MST.

Proof 3 *Assume e_1 is the MWOE from the MST fragment $F_a(V_a,E_a)$ and T is the MST. If $e_1 \notin T$, then there exists $e_2 \in T$ that is included in the cut $(V,V - V_a)$. Substituting e_1 for e_2 in T results in a spanning tree T' with less total weight as $w(e_1) < w(e_2)$ which means T was not minimum, meaning the original assumption $e_1 \notin T$ was wrong, therefore resulting in a contradiction.* $\square$

Lemma 3.2 Property 2
If weights of all of the edges of a connected graph are distinct, then the MST is unique.

Prim_MST divides the vertex set V of the graph G into two sets as P and Q where $P \cup Q = V$ at any time. Initially, the set P is initialized to an arbitrarily chosen vertex u and the algorithm always chooses the MWOE (u,v) from P, removes the vertex v from Q to P until all vertices are processed, as shown in Alg. 3.8.

Algorithm 3.8 *Prim_MST*

1: **Input** : $G(V,E,w)$
2: **Output** : $T \subseteq G$ which is minimal
3: $T \leftarrow \emptyset$
4: **select** $u \in V'$ arbitrarily
5: $P \leftarrow \{u\}; Q \leftarrow V \setminus \{u\}$
6: **while** $Q \neq \emptyset$ **do**
7:    **find** $MWOE$ (u,v) from vertices in P
8:    $P \leftarrow P \cup \{v\}$
9:    $Q \leftarrow Q \setminus \{v\}$
10:    $T \leftarrow T \cup \{(u,v)\}$
11: **end while**

The MST formed by *Prim_MST* algorithm is displayed in Figure 3.7 where the weights are shown next to the edges and the numbers in parentheses show the iteration of the *while* loop. Algorithm picks vertex 3 arbitrarily, then the first MWOE is edge (3,2) followed by (3,6), (3,7), (7,5) and (5,1) in the following steps. We could have elected the first vertex as the root of the MST in which case the predecessor of each vertex in the direction of the root can also be stored.

Prim_MST using adjacency matrix requires $O(n^2)$ time since we would need to run the *while* loop $O(n)$ times and we may need to perform $O(n)$ checks for each vertex. This time complexity can be reduced to $O(m \log n)$ steps using binary heap data structures. Kruskal's algorithm to construct an MST of a graph G works rather differently by first sorting the edges with respect to their weights and then including edges in the current MST fragment F in increasing order of their weights as long as they do not produce cycles in F. Implementing Kruskal's algorithm in the graph of Figure 3.7 would include edges (3,2),(7,5),(3,6),(3,7) and (5,1) in order and the resulting MST would be the same as found by the Prim's algorithm by *Property 2*, as the edge weights are distinct. Kruskal's algorithm requires $O(m \log m)$ steps to sort the edges which is $O(m \log n)$ time since $|E|$ is at most $|V|^2$ and $\log V^2 = 2 \log V$, therefore $O(\log E)$ is $O(\log V)$.

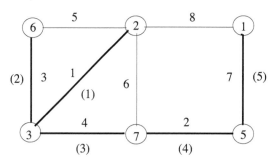

Figure 3.7: *Prim_MST* **example**

3.8 NP-complete Problems

All of the problems we have seen up to now have running times $O(n^k)$ for an input with size n where k is a positive constant. For example, finding the maximum value of an integer array (*Find_Max*) has $O(n)$ time complexity. Problems as such which can be solved in polynomial time are are called *tractable* and in class P meaning they have polynomial execution time. Problems that do not have a solution in polynomial time are called *intractable*. Some intractable problems can only be solved in exponential time and solutions to some problems do not exist at all.

The problems that we try to solve are most of the time either *optimization* problems where our aim is to find a solution with the best value (minimization or maximization of a certain parameter) or *decision* problems where we try to get an answer as *yes* or *no*. As an example, let us consider the *vertex cover* (VC) problem where given a graph $G(V,E)$, our aim is to find $V' \in V$ where every edge $e \in E$ is incident to at least one vertex in V'. The optimization problem associated with the VC problem asks to find the VC with the minimum number of vertices. The decision problem for VC searches an answer to the question whether there is a VC of size at most k for a given graph $G(V,E)$.

3.8.1 NP Completeness

NP-complete (nondeterministic polynomial time) problems are a subclass of intractable problems and no polynomial algorithms are known to solve them. However, the non-existence of such algorithms is not proven either. If any problem that is NP-complete (NPC) can be solved in polynomial time, then all of the NPC problems can be solved. Before investigating NPC problems any further, let us define what NP is. NP is a class of problems that can be verified in polynomial time by a *certifier* (verifier). A certifier inputs an instance I and a solution of a problem A and checks whether there is a solution for this instance.

Returning to the VC problem, given a graph $G(V,E)$, and the decision version of the VC problem, we need to check whether there is a certifier for this decision problem. Given an instance I of k vertices with solution V', we can write a simple certifier C where a check is done whether each $e \in E$ has an endpoint incident to a vertex in V' as shown in Alg. 3.9. V' has k elements and we need to check each vertex $v \in V'$ at the highest degree in the graph (Δ) times; therefore, the time complexity of the certificate algorithm is polynomial ($O(\Delta k)$). We can certainly assess after the check whether V' is a vertex cover or not; therefore, this problem is in NP. We strongly suspect there is a polynomial algorithm that finds a VC of size at most k. The brute force algorithm starts by checking each vertex whether a VC exists up to k vertex combinations for $\binom{n}{k}$ and hence has an exponential time.

Some important problems related to graph theory are as follows:

■ Independent Set Problem (IND): Given a simple, undirected graph $G(V,E)$, a set of vertices I is an independent set (IS) if there are no edges between any of these vertices. Formally, $\forall u, v \in I, (u,v) \notin E$.

Algorithm 3.9 *VC_Cert*

1: **Input** : $G(V,E), V'$ ▷ V' is the instance with k vertices
2: **Output** : YES or NO
3:
4: $S \leftarrow V'; P \leftarrow E$
5: **while** $S \neq \emptyset$ **do**
6:     **select** $u \in S$ arbitrarily
7:     **for all** $(u,v) \in E$ **do**
8:         $P \leftarrow P \setminus \{(u,v)\}$
9:         $S \leftarrow S \setminus \{u\}$
10:     **end for**
11: **end while**
12: **if** $P = \emptyset$ **then**   ▷ decide whether V' is a solution
13:     **return** YES
14: **else   return** NO
15: **end if**

  ■ *Optimization Problem*: Find an IS with the maximum size (MaxIS) in *G*.

  ■ *Decision Problem*: Is there an IS of size at least k in G where $k < n$?

■ Dominating Set Problem (DOM): Given a simple, undirected graph $G(V,E)$, a set of vertices D is a dominating set (DS) if any vertex is either an element of D or neighbor to a vertex in D. Formally, $\forall u \in (V - D), \exists v \in D : (u,v) \in E$.

  ■ *Optimization Problem*: Find a DS with the minimum size (MinDS) in *G*.

  ■ *Decision Problem*: Is there a DS of size at most k in G where $k < n$?

■ Clique Problem (CLIQUE): Given a simple, undirected graph $G(V,E)$, a set of vertices C is a clique if they are all connected to each other. Formally, $\forall u,v \in C, (u,v) \in E$.

  ■ *Optimization Problem*: Find the clique with the maximum size (Max-CLIQUE) in *G*.

  ■ *Decision Problem*: Is there a CLIQUE of size at least k in G where $k < n$?

■ Vertex Coloring Problem (COL): Given a simple, undirected graph $G(V,E)$, vertex coloring is assignment of colors to vertices such that no adjacent vertices have the same color. Formally, $\exists (c : V \rightarrow \{1,...,\sigma\} : \{(u,v) \in E \rightarrow c(u) \neq c(v)\}$

  ■ *Optimization Problem*: Find the minimum number of colors to color a graph *G*.

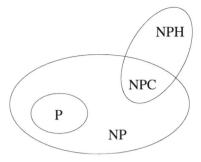

Figure 3.8: Relations between P, NP, NPC and NPH

- *Decision Problem*: Is there a coloring of *G* using at most *k* colors where $k < n$?

■ Hamiltonian Path Problem (HAM): Given a a simple, undirected graph $G(V, E)$, a Hamiltonian Path visits all of the vertices of *G* exactly once.

- *Optimization Problem*: Find a Hamiltonian Path in *G*.
- *Decision Problem*: Is there a Hamiltonian path in *G* of length *n*?

All of the problems in P are in NP as any problem that can be solved in polynomial time can also be solved by a certificate for an instance of it in polynomial time. We can therefore state $P \subseteq NP$. Whether P = NP is unknown but the opposite is widely believed. A problem is NPC if it is in NP and is as hard as any problem in NP. In order to show that a problem is as hard as any problem in NP, we need to show that it can be reduced to a problem in NP as described next, which means solving this problem is equivalent to solving the problem in NP. An NP-hard (NPH) problem is a problem which is as hard as any problem in NP. It should be noted that an NPH problem does not need to have a certificate.

Definition 1 *A problem A is NPC if it is both NPH and in NP, that is, NPC = NP ∩ NPH.*

The relations between these complexity classes are depicted in Figure 3.8.

3.8.2 Reductions

Reductions provide us with the means to identify similar problems. If a problem *A* can be reduced to another problem *B*, we can say that *A* is as hard as *B*. Formal definition of reduction is as follows.

Definition 2 *A problem A is polynomially reducible to another problem B if any input I_A of A can be transformed to an input I_B of B in polynomial time, and the*

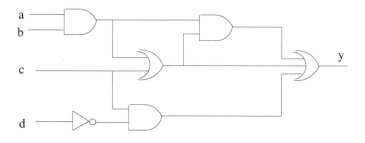

Figure 3.9: CSAT example

answer to input I_B for problem B is true if and only if the answer is true for problem A with the input I_A.

3.8.2.1 Satisfiability Problems

We will first define the circuit satisfiability problem (C-SAT) which is the first problem shown to be NPC. In this problem, we have a logic circuit consisting of AND, OR and NOT gates, with some unknown inputs and a single output; and are asked to find an instance of the input which yields a true or a false output value. The whole circuit has a single output as shown in Figure 3.9 where a, b, c, and d are the unknown inputs and y is the output. For n inputs, we would need 2^n time to check all possibilities. CSAT problem is in NP as we can verify in polynomial time a solution for a given set of inputs. Cook showed that CSAT is NP-hard [1], we can therefore state that CSAT is NPC.

Formula Satisfiability

In the formula satisfiability problem (SAT), we are given a Boolean formula and are asked to verify whether a given assignment to variables provides a true value for the formula. Given an instance of CSAT, we may transfer it to an equivalent logic formula. The logic circuit is satisfiable if and only if the equivalent formula is satisfiable. For the circuit of Figure 3.9, the output formula is:

$$(ab(ab \vee c)) \vee (ab \vee c) \vee (c\bar{d}) = ab \vee c \vee c\bar{d}$$

and is satisfiable with $a = 1$ and $b = 1$; $c = 1$; or $c = 1$ and $d = 0$.

Three Satisfiability

A Boolean formula is in conjunctive normal form (CNF) if it is a conjunction of clauses with each clause consisting of disjunction of several variables in true or negated form as follows:

$$(a \vee \bar{b} \vee c) \wedge (b \vee c) \wedge (\bar{a} \vee \bar{c})$$

The three satisfiability (3-SAT) problem is the formula satisfiability when the formula is restricted to exactly three literals in each clause. 3-SAT is NPC as C-SAT can be reduced to it by writing the output of the logic circuit in terms of logical variables and clearly, this can be performed in polynomial time.

3.8.2.2   3-SAT to Independent Set

In order to show the independent set (IND) problem is NPC, we need to show it is in NP and a known NPC problem can be reduced to IND problem. The input to the certificate consists of the vertex set V' which has k vertices. We can simply check each pair $(u, v) \in V'$ for any edge between them. This task can be performed in $O(k^2)$ time by investigating the adjacency matrix and we can determine if V' is an independent set or not, therefore $IND \in NP$. As the second step, we will show that 3-SAT can be reduced to IND problem by the following steps for a 3-CNF formula F of k clauses:

1. For each clause C in F, create a clause cluster of 3 vertices from the literals of C.

2. For each clause cluster (x_1, x_2, x_3) connect edges between all pairs of vertices in the cluster.

3. For each vertex x_i, connect edges between x_i and all of its complement vertices $\bar{x}_i$.

Once the graph G is formed, we claim F is satisfiable if and only if G contains an independent set of size k. As an example, given 3-CNF formula with size 4:

$$(a \vee \bar{b} \vee c) \wedge (\bar{a} \vee \bar{b} \vee c) \wedge (\bar{a} \vee b \vee c) \wedge (a \vee b \vee \bar{c}) \tag{3.8}$$

The graph G in Figure 3.10 is constructed using the above rules. Since G contains an IS of 4 shown with dark nodes $a, \bar{b}, c$ in (b), we conclude $IS \in NPC$ and a = 1, b = 0 and c = 1 input values satisfy Eqn. (3.8).

We can briefly sketch the correctness of our claim as follows. Each clause C_i of F must have at least one true literal v_i for satisfiability of F and let the set V' of G contain such vertices, one from each cluster. As all the vertices representing the cluster are connected, we can only take one vertex from each cluster. Also, since a variable and its complement cannot both be true, V cannot contain x_i and $\bar{x}_i$ at the same time which are connected. Based on this reasoning, V is an IS of G with size k. The transformation of 3-SAT to IND clearly can be performed in polynomial time which means 3-SAT can be reduced to IND problem. Since IND $\in$ NP, we conclude IND is NPC.

3.8.2.3   Independent Set to Vertex Cover

It can easily be noticed that, given a graph $G(V, E)$, a vertex set $S \in V$ is a vertex cover if and only if the set $V - S$ is an independent set of G. This situation is depicted in Figure 3.11. If $V - S$ was not an independent set of G, then there would be at least an

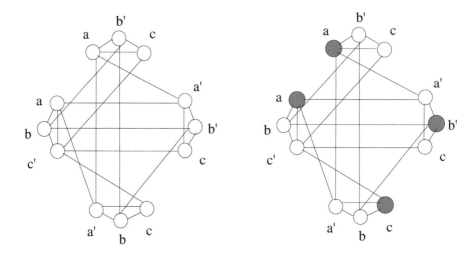

Figure 3.10: Reduction from 3-SAT to IND

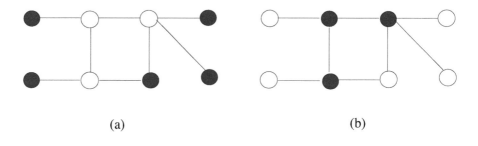

    (a)   (b)

Figure 3.11: Reduction from independent set to vertex cover. a) An independent set. b) A vertex cover from this independent set

edge $e \in E$ between two vertices in $V - S$ and therefore this edge will not be covered by a vertex in S and hence S will not be a vertex cover resulting in a contradiction. In order to solve an instance of IND with k vertices, we can simply check whether the remaining vertices form a vertex cover.

3.8.2.4 Independent Set to Clique

Given a graph $G(V, E)$, $S \in V$ is an independent set of G if and only if S is a clique in the complement graph G'. By definition, for any two vertices u and v to be in the independent set S, edge $(u, v) \notin E$ which means there will be edges between all vertex pairs $(u, v) \in S$ in G' resulting in a clique in G' with the same set of vertices. An example reduction is shown in Figure 3.12. We can therefore reduce IND to CLIQUE

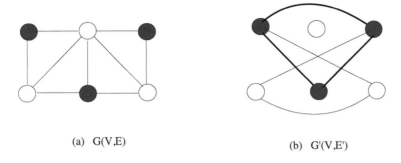

(a) G(V,E)　　　　　　　　　(b) G'(V,E')

Figure 3.12: Reduction from independent set to clique

by checking the proposed solution S in G whether IND is a clique in G'. Figure 3.13 displays the reduction relationships between the fundamental NPC problems.

Once a problem is found to be NPC, there is no need to search for a solution in P. The possible approaches for the solution are as follows:

■ We may use the exponential solution if there is one, for small values of input.

■ We can use heuristics which can lead to a solution in polynomial time. As an example, the heuristic to find the minimal VC could be always choosing the highest degree vertex in a graph and then deleting all of the edges around it. Since we need to cover as many edges as possible, this *highest degree first*

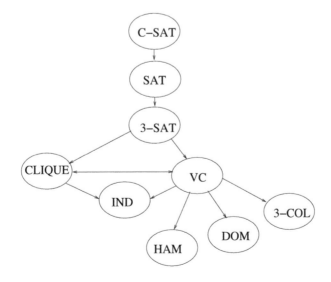

Figure 3.13: Diagram showing reductions between NPC problems

heuristic seems reasonable. We will however see that this is not a suitable heuristic. In general, a good heuristic provides sub-optimal results most of the time but there is no guarantee. Extensive testing is needed to show that a heuristic performs well in practice.

■ We may design an approximation algorithm that provides a sub-optimal solution at all times.

3.9 Coping with NP Completeness

When an NP-complete problem is encountered, we have various ways to tackle the problem. We will describe two relevant methods as applied to complex network problems, which can guide us in search of a solution to NP-complete problems in these networks. The first method involves limiting the search space by using some intelligent heuristics. This method usually referred to as *intelligent search* has two basic algorithm structures as *backtracking algorithms* and *branch and bound algorithms*.

3.9.1 Backtracking

A *search algorithm* finds an item with the required properties from a set of items. These algorithms are frequently used for searching in complex networks. The main idea of backtracking algorithms is to discard a solution to a given problem by the inspection of a smaller sub-solution. This method constructs a *state space tree* where each node represents a sub-state of its subtree. We start from the root state and proceed by evaluating cases along the branches of the tree. Whenever we encounter a falsified case, that is, a branch that does not satisfy the solution, we *backtrack* and check the other branches. We will illustrate this method by the use of a Boolean formula Ψ in CNF form as follows:

$$\Psi(a,b,c) = (a' \vee b') \wedge (a \vee b') \wedge (a' \vee c)$$

where a, b and c are boolean variables with *true* or *false* values only. We will grow a tree for the solution by starting with the logical values of any of the variables and if we encounter a branch that does not satisfy the Boolean formula, we will discard that branch of the tree and backtrack. We start with the variable a and test to see $a = 1$ does not violate the satisfiability of Ψ as shown in Figure 3.14. Proceeding with values of the variable b, true value for this variable when $a = 1$ results in false value of Ψ shown by X; therefore backtracking to the previous state in the tree is performed. Continuing in this manner, $a = 1$, $b = 0$ and $c = 1$ provide the solution required. The execution of this algorithm is shown by arrows and it can be seen that a number of recursive calls to the test procedure is done.

As can be seen, the backtracking algorithm is in fact a modified DFS of the solution tree. The pseudocode for a generic backtracking algorithm is shown in Alg. 3.10

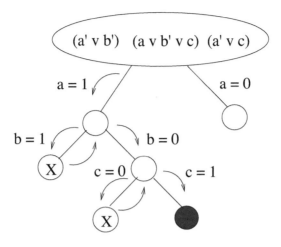

Figure 3.14: Backtracking algorithm example

where we search promising vertices of the tree and if a solution is found, it is output
and the algorithm stops. Otherwise, the leaves of the tree are searched recursively.

Algorithm 3.10 *Backtracking_Alg*

1: **procedure** BACKTRACK(v)
2:     **if** v is a promising node **then**
3:         **if** there is a solution at v **then**
4:             **output** the solution
5:             **stop**
6:         **else**
7:             **for all** u that is a child of v **do**
8:                 *Backtrack(u)*
9:             **end for**
10:         **end if**
11:     **end if**
12: **end procedure**

3.9.2 Branch and Bound

The *branch and bound* is an intelligent enumerating algorithm method and follows
a similar pattern to the backtracking algorithm method by using a state space tree. It
is used to find solutions to optimization problems. The branch and bound algorithms
record the best solution up to the search point, and they provide a bound on the best

value of the objective function that can be obtained by continuing with the current partial solution. If this value is not better than the best value found so far, that branch of the solution is abandoned.

In essence, branch and bound algorithms perform a BFS with pruning and enumerate all candidate solutions where some of these solutions are discarded by using upper and lower bounds of the quantity that is optimized [4]. The branch and bound algorithms consist of two steps as *branching* and *bounding*. The branching step recursively defines a tree branches of which are sub-solutions to the problem and a bound is computed for each node on the tree to decide whether it is promising or not. The main idea of this method is to avoid growing all of the branches of this tree by carefully selecting branches that are more promising than others. The bound of the best value of the objective function by growing a branch is estimated and if this value does not meet the requirements, that branch is discarded in the bounding step. The pseudocode for generic branch and bound algorithms is displayed in Alg. 3.11.

Algorithm 3.11 *Branch_and_Bound*

1: **procedure** BRANCHBOUND(T,v, *soln*)
2:    T is the solution space tree; $v \in T$
3:    *soln* : best solution found so far
4:    **if** v provides a more optimum solution than *soln* **then**
5:       *soln* $\leftarrow v$
6:    **end if**
7:    **generate** the children set C of v
8:    **compute** bounds for each $c_i \in C$
9:    **select** $C' \in C$ which have promising upper bounds
10:   **for all** $c_i \in C'$ **do**
11:      *BranchBound*(T,c_i, *soln*)
12:   **end for**
13: **end procedure**

3.10   Approximation Algorithms

In optimization problems, we are usually asked to minimize or maximize a parameter. For an instance I of an optimization problem, we can define OPT(I) as the value of the optimum cost of solution for an instance I. If $A(I)$ is the algorithm under consideration, then the approximation ratio $\alpha(I)$ of this algorithm is defined as:

$$\alpha_A = max_I \frac{A(I)}{OPT(I)} \tag{3.9}$$

For minimization problems, $\alpha \geq 1$, however, for maximization problems we need to take the reciprocal to have $\alpha \geq 1$. Let us illustrate this concept for the VC problem where we are asked to find the minimum number of vertices of a graph such that at least one end point of every edge of graph is incident to at least one of the vertices of this set. For the MinVC problem mentioned, we can design an approximation algorithm called *Approx_MVC*. This algorithm randomly selects an unassigned edge from the graph and includes the end points of this edge in MVC and deletes all edges incident to these end points from the edge set as shown in Alg. 3.12. Removing an edge from the graph is continued until there are no edges left which means all edges are covered by the included end points of the selected vertices, therefore these vertices constitute a vertex cover.

Algorithm 3.12 *Approx_MVC*

1: Input $G(V, E)$
2: $S \leftarrow E, MVC \leftarrow \emptyset$
3: **while** $S \neq \emptyset$ **do**
4:     **pick** any $(u, v) \in S$
5:     $MVC \leftarrow MVC \cup \{u, v\}$
6:     **delete** all edges incident to either u or v from S
7: **end while**

The execution of this algorithm in a graph of 8 vertices labeled 1,...,8 is shown in Figure 3.15. The first edge randomly selected is $(6, 8)$ and the edges $(5,8)$, $(2,8)$, $(4,8)$, $(5,6)$ and $(4,6)$ incident to vertices 6 and 8 are deleted from the graph resulting in the graph of (b). The next randomly selected edge is $(2,3)$ which results in the inclusion of vertices 2 and 3 in the cover and covering of the edges $(2,7)$, $(3,7)$ and $(3,1)$. The finally selected edge $(4,7)$ completes the algorithm, resulting in a minimal cover consisting of vertices 8, 2, 3, 6, 4, and 7 as shown in (d) and the size of this vertex cover is 6. A MinVC for this graph has vertices 8, 3, 6, and 7 with a size 4. For this example implementation, the *Approx_MVC* algorithm has provided 1.5 approximation to MinVC.

Analysis

Theorem 3.2
Alg. 3.12 provides a vertex cover in $O(m)$ time and the size of MVC is $2 |MinVC|$.

Proof 4 *Since the algorithm continues until there are no more edges left, every edge is covered; therefore the output from Seq1_MVC is a MVC, taking $O(m)$ time. The set of edges picked by this algorithm is a matching, as edges chosen are disjoint and it is maximal as addition of another edge is not possible. Since two vertices are covered for each matched edge, the approximation ratio for this algorithm is 2.* □

A simple procedure to find a MVC would then consist of finding a maximal matching of the graph G in the first step and including in the maximal vertex cover

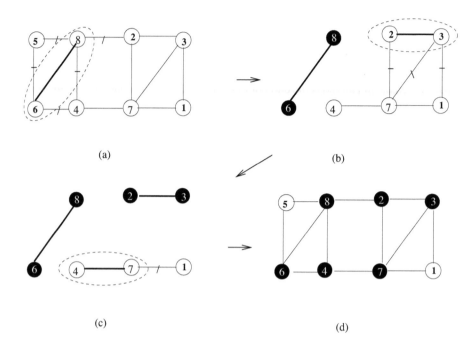

(a)

(b)

(c)

(d)

Figure 3.15: Approximate vertex cover example

all of the end points of the edges found in the first step. The size of a maximal matching of G determines the lower bound of the MVC as the vertex cover cannot have a smaller magnitude than this size; therefore any minimal vertex cover of G must be at least as large of the size of a maximal matching of G. The following observation can then be made; finding a maximal matching of small size will result in a small size vertex cover when maximal matching is used to find a minimal vertex cover.

We may try a different algorithm and select a vertex that has at least one uncovered edge incident to it and include it in the cover at each step. This algorithm will work poorly for the case of a star graph for example. Searching for another algorithm, it may seem using a a greedy approach where we always choose the vertex with the highest degree should provide improvement as we will be attempting to cover more edges than a random selection on the average. We start with the highest degree vertex first, include it in the cover and remove all edges incident to it from the graph and continue until no more edges are left as shown in Alg. 3.13.

The operation of this heuristic is shown in Figure 3.16 with the same example and it can be seen that the result is a MVC with only 4 vertices instead of 6 with the random edge choosing heuristic. Unfortunately, this heuristic does not always improve the heuristic that randomly chooses an edge. It can be shown that it does not even have a constant bound and its ratio bound grows as $\theta(\log n)$.

Algorithm 3.13 *Greedy_VC*

1: Input $G(V,E)$
2: $V' \leftarrow V, E' \leftarrow E, MVC \leftarrow \emptyset$
3: **while** $E' \neq \emptyset$ **do**
4:     **pick** the highest degree vertex $v \in V'$
5:     $V' \leftarrow V'\{v\}$
6:     $MVC \leftarrow MVC \cup \{v\}$
7:     **remove** all edges incident to v from E
8: **end while**

3.11 Parallel Algorithms

Certain computing applications such as image processing, large matrix operations and operations on large graphs representing complex networks require computational power which may not be available by one computational element. A parallel algorithm executes on many different processing elements providing a *speedup* in the total processing time of an algorithm. This speedup is usually expressed as the ratio of the time taken for a sequential algorithm to solve the problem to the execution time of the parallel algorithm. Parallel algorithms may be designed by exploiting

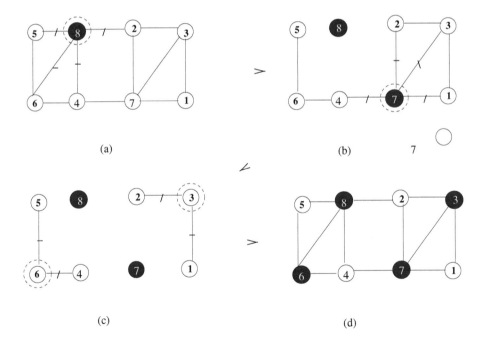

(a)

(b)

7

(c)

(d)

Figure 3.16: The greedy heuristic

parallelism of parts of an already existing sequential algorithm or by designing a completely new parallel algorithm.

3.11.1 Architectural Constraints

Important constraints that should be considered are the single or multiple instruction control; shared memory or distributed memory architectures and input/output models. In a single processor computer, there is one instruction executed at a time. Parallel computers with m processors however, can execute m instructions concurrently. Parallel computing architectures where all processors execute the same instruction on different data synchronously are called single-instruction-multiple-data (SIMD) architectures. In multiple-instruction-multiple-data (MIMD) architectures, different instructions usually on different data are executed. MIMD architectures are more general and can easily be configured by connecting various computational nodes over a communication network whereas SIMD architectures are more specifically designed and used to perform same operation on large data sets such as very large matrices. A single processor computation element is called single-instruction-single-data (SISD) in this classification.

In the *shared-memory model*, all processors can access a common RAM memory for communication and parallel computing architectures using shared memory are called parallel random access machines (PRAMs). Mechanisms to monitor access to shared memory are needed in the PRAM model. PRAM architecture can be implemented using the following models:

1. Exclusive-read, exclusive-write (EREW) model

2. Concurrent-read, exclusive-write (CREW) model

3. Concurrent-read, concurrent-write (CRCW) model

EREW model is the most commonly used model, however, access to shared memory should be monitored by locks or other structures provided by the operating system. Distributed memory parallel computers have the advantage that communication and synchronization are provided by the transfer of messages only.

3.11.2 Example Algorithms

As a first example which uses the PRAM-EREW model, we will describe an algorithm to find the sum of the elements of an array $a[1, ..., n]$ holding real values. The sequential algorithm for this purpose will simply perform n additions of the values of array to the variable *sum*. The parallel EREW algorithm executed by a processor P_i, $1 \le i \le k$ finds only the partial sum of the entries dependent on its identity as shown in Alg. 3.14. We have assumed each parallel process knows its identity, the number of total processes k, and the start address m_1 which contains the size of of the array followed by the location where the total sum will be stored. Since we need to provide exclusive read and write operations, the shared memory area is protected by the lock variables provided by the operating system.

Algorithm 3.14 *Parallel_Sum Code for P_i*

1: **Input** : address m_1
2: **Output** : sum *my_sum*
3:    $size \leftarrow [m1]$ ▷ code for process P_i
4:    $my\_start \leftarrow (i * size/k) + 1$
5: **set lock**
6: **for** $x = 1$ to $size/k$ **do**   ▷ copy my part from shared memory
7:      $my\_array[x] \leftarrow [my\_start + x]$
8: **end for**
9: **unlock**
10: $my\_sum \leftarrow 0$
11: **for** $x = 1$ to $size/k$ **do**   ▷ calculate my sum
12:      $my\_sum \leftarrow my\_sum + my\_array[x]$
13: **end for**
14: **set lock**
15: $[m_1 + 1] \leftarrow my\_sum$ ▷ store my sum in the location $m_1 + 1$
16: **unlock**

As a concrete example of MIMD architecture, we will show the calculation of PI which is given by:

$$PI = \int_0^1 \frac{4}{1+x^2} \tag{3.10}$$

which is the area under the curve $(4/(1+x^2))$ between the x coordinates 0 and 1 as shown in Figure 3.17.

As an attempt to calculate PI in parallel using N processors, we can divide the area under this curve to N slices and have each processor P_i calculate its own slice of area. The striped area under the curve is the output of the parallel processing and it can be seen that using many processing elements will result in more precision. We assume that each parallel process has a distinct identifier $1 \leq i \leq k$, and we assume $k = N$ such that each processor computes the area of one slice. One of the processors called the *initiator* which is not involved in the calculation can first start parallel processing by sending the *worker* nodes the number of partitions and then collecting all the results to find the total area. Alg. 3.15 shows the code for the initiator and Alg. 3.16 shows the code executed by the parallel processing nodes. The reception of the subareas from the nodes is synchronized with their sending by the underlying parallel computing software platform. One such widely used parallel processing software platform is message passing interface (MPI) which provides the communication routines between the parallel processing nodes in a variety of forms [9].

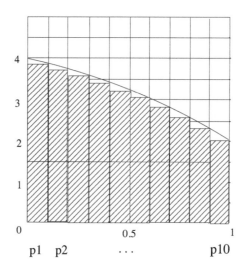

Figure 3.17: Parallel calculation of PI

Algorithm 3.15 *Calc_PI Initiator*

1: **Output** : *total_area*
2: **for** $i = 1$ to N **do**
3:     **send**(i, N) to P_i
4: **end for**
5: **for** $i = 1$ to N **do**
6:     **receive**$(area)$ from P_i
7:     *total_area* $\leftarrow$ *total_area* + *area*
8: **end for**

Algorithm 3.16 *Calc_PI , node P_i*

1: **Input** : identity i, number of partitions N
2: **Output** : area calculated *my_area*
3: **receive**(my_id, N) from *initiator*
4: *width* $\leftarrow 1/N$
5: *x_coord* $\leftarrow my\_id * width$
6: *y_coord* $\leftarrow 4/(1 + x\_coord^2)$
7: *my_area* $\leftarrow x\_coord * y\_coord$
8: **send**(my_area) to *initiator*

There may be improvements to this algorithm where the initiator could also be involved in the calculation and the worker processes calculate more than one slice (see Ex. 7). The most important benefit of using a parallel algorithm is the speedup achieved with respect to a sequential algorithm to perform the same task and this is needed especially for time consuming tasks.

3.12 Distributed Systems and Algorithms

A distributed system consists of computational nodes connected over a network which cooperate to achieve a common goal. Mobile ad hoc networks, the grid and wireless sensor networks are examples of the distributed systems. Routing of the packets using lowest cost available paths is an example of a common task performed by the nodes of a distributed system. However, since no node has a complete view of the network, and the states of the nodes in the network, accomplishing common tasks becomes much more difficult than executing the algorithm sequentially as in the routing task example. Although the designer of a distributed algorithm is faced with many challenges such as the synchronization of algorithms running on different nodes of the network, distributed systems have certain advantages. One such benefit is *resource sharing* where database or peripherals over a network can be shared by a number of users. Provision of fault tolerance is another important benefit of distributed systems where computational resources and software can be replicated and in the event of the failure of the active software, a replica can resume processing. Even when these benefits are not so clearly visible, there are certain applications such as the airline routing systems and distributed control systems where the use of distributed systems is inevitable.

A distributed algorithm executes on the nodes of the distributed systems. A *symmetric* distributed algorithm has the same code running at each node with possibly different data. An asymmetric distributed algorithm on the other hand, consists of different codes running on different nodes. Another important issue is the synchronization of distributed algorithms at different nodes. In *synchronous algorithms*, execution is performed in synchronous rounds started by a special process called the *initiator* in the network. Each round when started, consists of sending of messages to the neighbors; reception of messages from the neighbors and performing some computation using the received data. In asynchronous algorithms, there is no central control of execution, however, there may still be a single initiator which controls message transfer operations. Distributed algorithms started by more than one initiators at different nodes are called concurrent-initiator algorithms. Based on these two different levels of specifications, a distributed algorithm may be a:

- Synchronous single-initiator algorithm (SSI)

- Asynchronous single-initiator algorithm (ASI)

- Synchronous concurrent-initiator algorithm (SCI)

- Asynchronous concurrent-initiator algorithm (ACI)

ASI algorithms are simpler in design but termination condition should be carefully considered to prevent endless running of the algorithm. In SCI algorithms, there are more than one initiators which need to be synchronized requiring possibly hardware synchronization such as clock ticks. ACI algorithms provide more flexibility than others but are more difficult to design and analyze due to uncertainties involved. SSI algorithms, in general are easier to analyze and are frequently used.

The order of execution in any type of distributed algorithm is influenced greatly by the type of message received from neighbors and depending on this type, a different action is performed. Alg. 3.17 shows a template for a typical distributed algorithm where the *while* loop is executed until the value of the *flag* variable becomes *true* [3].

Algorithm 3.17 *General Distributed Algorithm Structure*

1: **int** i, j ▷ i is this node; j is the sender of the current message
2: **while** $\neg finished$ **do**   ▷ all nodes execute the same code
3:      **receive** $msg(j)$
4:      **case** $msg(j).type$ **of**
5:              $\underline{msg\_type\_1}$: $Action\_1$
6:              ... : ...
7:              $\underline{msg\_type\_n}$: $Action\_n$
8:      **if** $condition$ **then**
9:          $finished \leftarrow true$
10:     **end if**
11: **end while**

A template of a SSI algorithm is shown in Alg. 3.18 where a round k is started by the $round(k)$ message from the initiator to all nodes. This message is typically transmitted from parents to children over a spanning tree which is built prior to the algorithm execution.

Upon receiving this message, each node sends its results from the previous round $(k-1)$ to all of its neighbors and receives results of the previous round from all of its neighbors. This ordering of messages is not strict however. We could have nodes receiving messages from neighbors about the results of a previous round, performing some computation using these results and then sending results to the neighbors. In order to signal the end of the round, each node sends an *upcast* message to its parent upon completion of its task. When a parent receives *upcast* messages from all of its children and itself finishes, it sends an *upcast* message to its parent. This way, the initiator finds that the round is over when all upcast messages are received from its children. Our focus in this book is mainly concerned with distributed algorithms in complex networks such as MANETs and WSNs and we will be using the SSI model more frequently than the others.

Algorithm 3.18 *Sample_SSI*

 1: **boolean** *finished, round_over* ← *false*
 2: **message type** *round, info, upcast*
 3: **while** ¬*round_over* **do**
 4:     **receive** *msg(j)*
 5:     **case** *msg(j).type* **of**
 6:             round(k) :   **send** *my_result(i, k − 1)* to all neighbors
 7:                              **receive** *results(k − 1)* from all neighbors
 8:                              **do** some computation, *finished* ← *true*
 9:             upcast   :   **if** *upcast* received from all children **and** *finished* **then**
10:                                  **send** *upcast* to *parent*
11:                                  *round_over* ← *true, finished* ← *false*
12: **end while**

3.13   Chapter Notes

In this chapter, we first analyzed the complexity of algorithms and reviewed fundamental algorithm methods with representative examples. Greedy algorithms look for local optimal solutions only and are suitable only for certain applications such as the MST problem. We saw important graph algorithms such as BFS and DFS and we will look into more graph algorithms for clustering in Chapter 8. Divide and conquer algorithms divide a problem into smaller subproblems which are solved and the solutions are then combined to yield the final solution. Dynamic programming method also divides a problem into smaller subproblems and the solutions that are found are stored to be be reused. This method is frequently used for bioinformatics problems.

We then described problems that do not have solutions in polynomial time. For these problems, we have a number of methods to search for solutions in polynomial time. Backtracking which performs a DFS, and branch and bound method which uses BFS to search the state space tree are the two methods of intelligent search to reduce the solution search space of a given problem. We have seen that approximation algorithms with sub-optimal performances can be used for these hard problems called NP-complete problems. Parallel algorithms are used when performance is the key issue such as needed in dense computations. For network oriented problems where nodes of a network cooperate to accomplish an overall task and should be aware of the result of this task, distributed algorithms are usually the only choice.

A comprehensive introduction to algorithms can be found in [5] and in [33]. Parallel algorithms with implementations are described in [6]. Distributed algorithms in a general view are analyzed in [8] and [10] and distributed graph algorithms for computer networks are described in [3].

Exercises

1. Show step-by-step execution of *Mergesort* algorithm in the array {12,1,8,9,7,3,6,5}.

2. Show the execution of BFS algorithm in the sample graph shown in Figure 3.18.

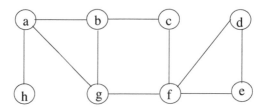

Figure 3.18: Example graph for Ex. 2 and Ex. 8

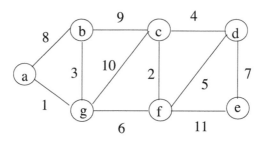

Figure 3.19: Example graph for Ex. 3

3. Find the MST of the graph shown in Figure 3.19 using both Prim's and Kruskal's algorithms and show that they are the same since there is a unique MST because of distinct edge weights.

4. Reduce the following 3-CNF formula to IND and determine if there is a combination of input variables that yields a true value output.

$$(a \vee \bar{b} \vee \bar{c}) \wedge (\bar{a} \vee \bar{b} \vee c) \wedge (a \vee b \vee c) \wedge (\bar{a} \vee b \vee \bar{c})$$

5. An independent set is shown by dark vertices in the graph in Figure 3.20. Show that IND problem can be reduced to CLIQUE and VC problems using this graph.

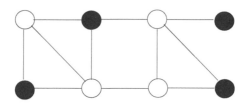

Figure 3.20: Example graph for Ex. 6

6. Provide the pseudocodes of parallel algorithms for the initiator and the worker nodes to perform a matrix multiplication of two $n \times n$ matrices so that A = B $\times$ C is formed at the initiator node. There are k parallel processing elements and the initiator sends matrix C and n/k rows of matrix B to each processor.

7. Improve the parallel algorithm to compute PI by having also the initiator performing calculation and each worker calculating a number of slices for better precision.

8. Each node in a distributed system has a unique identifier. Write the pseudocode for symmetric ASI distributed algorithm running at the nodes of a distributed system which enables the identifers of all nodes to be stored at each node at the end of the execution. Show the execution of this algorithm in the example graph of Figure 3.18.

References

[1] S. Cook. The complexity of theorem proving procedures. In *Proceedings of the Third Annual ACM Symposium on Theory of Computing*. pages 151-158, 1971.

[2] Cormen et al., Introduction to Algorithms, 3rd Ed., The MIT Press, 2009.

[3] K. Erciyes. *Distributed Graph Algorithms for Computer Networks*. Computer Communications and Networks Series, Springer, ISBN 978-1-4471-5172-2, 69-70, 2013.

[4] A. H. Land and A. G. Doig . An automatic method of solving discrete programming problems. *Econometrica*, 28(3):497-520, 1960.

[5] S. Skiena. *The Algorithm Design Manual*. Springer, ISBN-10: 1849967202, 2008.

[6] A. Grama, G. Karypis, V. Kumar, and A. Gupta. *Introduction to Parallel Computing*, Addison-Wesley, 2nd Ed., ISBN-10: 0201648652, 2003.

[7] A. Levitin. *Introduction to the Design and Analysis of Algorithms*. Pearson International Ed., 3rd Ed., ISBN: 0-321-36413-9, 2011.

[8] N. A. Lynch. *Distributed algorithms*. The Morgan Kaufmann Series in Data Management Systems, 1996.

[9] M. Snir, J. Dongarra, J. S. Kawalik, S. Huss-Lederman, S. W. Otto, and D. W. Walker. *MPI: The Complete Reference*. MIT Press, 1998.

[10] G. Tel. *Introduction to distributed algorithms*, 2nd Ed., Cambridge University Press, Cambridge, 2000.

Chapter 4

Analysis of Complex Networks

4.1 Introduction

We have seen that graphs can be conveniently used to model complex networks. Our next step would be the investigation of whether we can obtain some properties of complex networks using their graph representations. Before starting with this investigation, determination of what properties to search would probably be a good starting point. We may want to know which nodes of the complex network are more important than others, or on a larger scale, which groups of nodes are more closely related to each other than to the rest of the network such as a social network. It may also be interesting to see if some subgraph pattern is repeating itself significantly in the network which may indicate a fundamental network functionality.

Our analysis of complex networks in this chapter is restricted to the topological features of these networks to investigate the properties such as above. Graph theory with its rich background may not be adequate for our purpose and we may need to develop new concepts and metrics to be able to determine certain properties of the complex network we are investigating.

We start this chapter with a simple graph property, the degrees of vertices and see that it may give us a lot of information about the structure and furthermore, the type of the complex network. We then define a new parameter called the *clustering coefficient* which gives us hints about the closely related groups of nodes and the density of the graph in the network. Node and edge centralities provide us with information about the criticality of nodes and edges in the network and are briefly introduced in this chapter. We also describe network motifs which are subgraph patterns statistically appearing more frequently than other subgraph patterns and may provide insight to the overall functioning of the complex network.

4.2 Vertex Degrees

The degree of a vertex in a graph is the number of edges incident to it. The degree is a simple yet effective parameter for the analysis of complex networks. The following theorem relates the sum of degrees of a graph to the number of its edges.

Theorem 4.1
For any graph G(V,E), the sum of the degrees of vertices is twice the number of its edges stated formally as follows:

$$\sum_{v \in V} \delta(v) = 2m \tag{4.1}$$

where $\delta(v)$ is the degree of vertex v.

Proof is trivial as we count each edge twice for both of its endpoints and the total count is twice the number of edges. There is a corollary to this theorem as follows:

Corollary 5 *For any graph G, the number of vertices with odd degree is even.*

Proof 6 *We can divide the vertices of G into V_o for odd degree vertices and V_e for even degree vertices. Then,*

$$\sum_{v \in V} \delta(v) = \sum_{u \in V_e} \delta(u) + \sum_{w \in V_o} \delta(w) \tag{4.2}$$

As the left hand side of this equation and the first term at the right are even, the last term which is the sum of the degrees of all odd degree vertices should be even. We conclude the number of odd degree vertices in any graph should be even.

The average degree of a graph is the average value of the sum of all degrees and combining with Eqn. (4.1), we get:

$$\delta(av) = \frac{1}{n} \sum_{v \in V} \delta(v) = \frac{2m}{n} \tag{4.3}$$

Definition 4.1 density   The *density* $\rho(G)$ of a graph $G(V,E)$ is the ratio of the number of its edges to the number of edges of a fully connected graph with the same order and combining with Eqn. (4.3) yields:

$$\rho(G) = \frac{2m}{n(n-1)} = \frac{\delta(av)}{(n-1)} \tag{4.4}$$

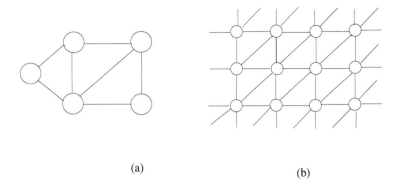

(a)

(b)

Figure 4.1: Degree sequence of graphs

For large networks where $n \gg 1$, $\rho = \delta(av)/n$. If ρ does not change significantly as $n \to \infty$, the network is called *dense*. On the other hand, if $\rho \to 0$ as $n \to \infty$, the network is called *sparse*. Graphs in which all vertices have the same degree are called *regular graphs* and in a *k-regular* graph, all vertices have the same degree of k. Figure 4.1.a displays a graph that has an average degree of 2.8 with $\rho = 0.7$ and a 6-regular network is shown in (b).

The *degree variance* $\sigma(G)$ of a graph $G(V,E)$ is defined as the average of the square of the distances of degrees to the average degree as follows:

$$\sigma(G) = \frac{1}{n-1} \sum_{v \in V} (\delta(v) - \delta(av))^2 \qquad (4.5)$$

and the *mean* of absolute distance between node degrees and the average degree of a graph G is given by:

$$\tau(G) = \frac{1}{n} \sum_{v \in V} |\delta(v) - \delta(av)| \qquad (4.6)$$

4.2.1 Degree Sequence

The *degree sequence* of a graph G is the listing of the degrees of its vertices, usually in descending order. The sum of this list should be an even number by Theorem 4.1 for simple graphs. If the given degree sequence represents a simple graph, that is, a graph with no loops or multi edges, the list is called *graphic*. The degree sequence of the graph in Figure 4.1.a is $\{4,3,3,2,2\}$. In *regular graphs* such as the one in Figure 4.1.b, each vertex has the same degree. Based on Theorem 4.1, it is not difficult to check whether a degree sequence is a valid one. For example, the degree sequence $\{6,5,4,3,1\}$ could not be valid as the sum is odd.

Two isomorphic graphs have the same degree sequence but two graphs with the same degree sequence may not be isomorphic as shown in the graphs of Figure 4.2

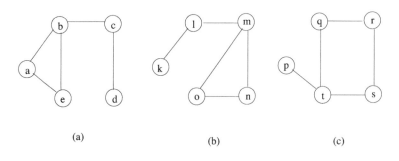

(a) (b) (c)

Figure 4.2: Graph isomorphism. Graphs in (a) and (b) are isomorphic but (a) and (c), and (b) and (c) are not isomorphic

in which all graphs have the same degree sequence of $\{3,2,2,2,1\}$ but they are not all isomorphic. We can therefore state that for two graphs G_1 and G_2 to be isomorphic, to have the same degree sequence is a necessary but not a sufficient condition. We will in see in Chapter 11 that checking whether two graphs are isomorphic is difficult and an NP-hard problem.

4.2.2 Degree Distribution

Plotting the degree sequence of a graph against the magnitude of degrees provides us with its *degree distribution*. The degree distribution $P(k)$ of an undirected graph G is the probability that any randomly chosen vertex has a degree k. It can formally be defined as follows.

Definition 3 *The degree distribution $P(k)$ of degree k in a graph G is given as the fraction of vertices with the same degree to the total number of vertices as below.*

$$P(k) = \frac{n_k}{n} \tag{4.7}$$

where n_k is the number of vertices with degree k. For a directed graph, it is the probability that a randomly chosen vertex has in-degree $P(k_{in})$ and out-degree $P(k_{out})$ respectively. In the case of random networks, degree distribution follows binomial distribution which approaches Poisson in the limit for large n values. In these networks, the degrees of the nodes are distributed around the average node degree. In *scale-free networks* however, few nodes called *hubs* have much higher degrees than the rest of the nodes, and the degree distribution displays a heavy tailed curve. In these networks, the degree distribution follows the power law $P(k) \sim k^{-\gamma}$ where the number of nodes having the degree decreases greatly as the degree increases. The degree distributions of these networks are shown in Figure 4.3.

The cumulative degree distribution $P_c(k)$ is defined as the probability of a randomly chosen vertex having a degree greater than k. The cumulative degree distribution $P_c(k) \sim k^{1-\gamma}$ when the degree distribution follows the power law [6].

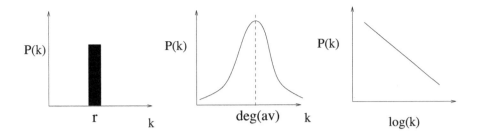

Figure 4.3: Degree distribution of graphs. a) The degree distribution of a homogeneous graph where all vertices have degree *r*. b) The degree distribution of a random graph. c) The degree distribution of a scale-free network such as the world wide web

We can further investigate the correlations between the degrees of neighboring nodes of the network. Based on these correlations, we can classify complex networks as *assortative* or *disassortative*. In assortative complex networks, high degree nodes are commonly connected to each other as in social networks where popular persons are usually friends with each other. The dissassortative networks manifest high degrees nodes connected to low degree nodes as in biological and technical networks [6].

4.3 Communities

Real-world complex networks often contain closely related group of nodes as subgraphs. *Communities* or *clusters* in a network are the groups of nodes that have higher connections among them than the rest of the nodes in the network. Figure 4.4

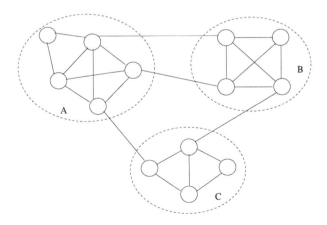

Figure 4.4: Clusters in a network

displays a network with clusters *A*, *B* and *C*, and nodes in each cluster are more closely related to each other than to the other nodes, as shown. Each cluster may be representing close friends in a social network where an edge between two nodes indicates a relationship between them or these clusters may be representing the closely interacting proteins in a PPI network and the proteins in a cluster may be involved in an important function. Detecting such communities or clusters has many implications in social networks, biological networks and technological networks, and is a fundamental research area in complex networks as well as in computer science. Existence of clusters in a network and its density can be investigated using the clustering coefficient parameter which is described in the next section.

4.3.1 Clustering Coefficient

Given a graph $G(V,E)$, the *clustering coefficient* of a node $v \in V$ shows how well connected the neighbors of v are. It can be formally defined as follows.

Definition 4 *Given a graph is $G(V,E)$ where V is the set of its vertices and E is the set of its edges, the clustering coefficient $cc(v)$ of a vertex $v \in V$ is defined as the ratio of the existing connections between its neighbors to the number of maximum possible connections between them as follows:*

$$cc(v) = \frac{2m_v}{n_v(n_v - 1)} \tag{4.8}$$

where n_v is the count of the neighbors of vertex v, and m_v is the existing number of edges between these neighbors.

The clustering coefficient is sometimes called the *local clustering coefficient*. The average clustering coefficient $CC(G)$ of a graph G is the average value of the clustering coefficient values of the whole networks as follows:

$$CC(G) = \frac{1}{n} \sum_{v \in V} cc(v), \tag{4.9}$$

where n is the number of nodes. A low average CC implies low connectivity between pairs of nodes. In other words, $CC(G)$ of a graph shows how dense it is. Figure 4.5.a displays a graph with an average CC of 0.40 and another graph with the same number of nodes but more connected with a higher CC of 0.63 is shown in (b) which is more connected and denser.

A different method to calculate clustering coefficients was proposed by Watts et al. [11] by the use of triangles and triads. A *triangle* is a complete graph with three vertices and a vertex v is in a triangle if it is one of the vertices of this triangle. A *triad* of a vertex v is a linear graph of three vertices where v is in the middle. It can be seen that the number of triangles that a vertex v is incident is the number of existing connections between its neighbors, and the number of triads it is in the middle is the maximum possible number of connections between its neighbors. Based on this reasoning, we can state that the local clustering coefficient $cc(v)$ of a vertex v is the

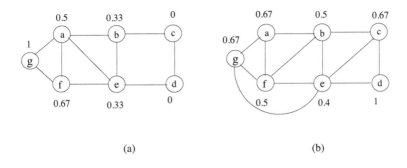

(a)   (b)

Figure 4.5: Clustering coefficients of graphs

Table 4.1: Clustering Coefficient Calculation

	a	b	c	d	e	f	g
n_t	3	1	0	0	2	2	1
n_x	6	3	1	1	6	3	1
cc	0.5	0.33	0	0	0.33	0.67	1

---

ratio of the number of triangles it is included, to the number of triplexes it is incident in the middle as follows:

$$cc(v) = \frac{n_t(v)}{n_x(v)} \text{ where } n_x(v) = \binom{\delta(v)}{2}, \tag{4.10}$$

and $n_t(v)$ and $n_x(v)$ denote the number of triangles and triads of the vertex v respectively. Applying this procedure to the vertices of the graph in Figure 4.5.a yields the values in Table 4.1. Clustering coefficients calculated using this method should give the same values using Eqn. (4.9).

Based on these metrics, the network transitivity $\tau(G)$ can be defined as follows:

Definition 5 *Given a simple connected graph $G(V,E)$ with $n_t(G)$ distinct triangles and $n_t(G)$ distinct triads, network transitivity $\tau(G)$ is the ratio of $n_t(G)$ to $n_x(G)$, by considering each triangle that a vertex is incident should be counted three times in the graph:*

$$\tau(G) = \frac{3n_t(G)}{n_x(G)} = \frac{\sum n_t(v)}{\sum n_x(v)} \tag{4.11}$$

The network transitivity $\tau(G)$ and the average network clustering coefficient $CC(G)$ are closely related as each is dependent on the number of triangles and triads but they are not equal. The clustering coefficient emphasizes weights of the low degree nodes, while the transitivity ratio considers the weights of the high degree nodes more.

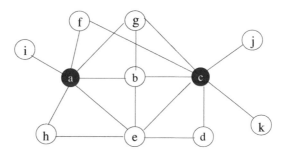

Figure 4.6: The matching index

4.4 The Matching Index

Two unconnected nodes of a complex network may have similar functionality and in order to measure node similarity to some extent, *matching index* is defined. In normalized form, this index evaluates the similarity between two vertices by finding the ratio of the number of their common neighbors to their total neighbor number [7]. An example graph is shown in Figure 4.6 where the matching index between the vertices a and c is to be evaluated. Their common neighbors are $\{f, g, b, e\}$ for a total of four vertices. The neighbors of a are $\{f, g, b, e, h, i\}$ and c has $\{f, g, b, e, d, j, k\}$ as neighbors. The union of these two neighbor sets has 9 distinct vertices, therefore, the matching index $M_{a,c}$ between the vertices a and c is 4/9 = 0.44.

The matching index can be computed for three or more vertices in which case we need to find common neighbors and total distinct neighbors of a set of vertices [11]. We propose here that it can also be evaluated for common k-hop neighbors which are common neighbors within k-hop distance from two or more vertices. We can also define two or more vertices similar if their distances to all other vertices in the graph are approximately equal [12, 7].

4.5 Centrality

Centrality is a measure of the importance of a node or an edge in a complex network. In this method, we attempt to assign an importance value to each node or an edge based on its topological position in the network. In the simplest form, *degree centrality* is the degree of a node. For example, a person with many acquaintances is more influential than a person with fewer links to others in a social network. The most influential nodes in the graph of Figure 4.5.a are the nodes a and e as they have the highest degree.

For practical purposes, we may be more interested to find nodes that can be reached using shortest paths from any nodes in the network. *Closeness centrality* attaches importance to nodes based on their average distances to all other nodes in the network. A node with an average shorter distance to all others is considered more important than others in this case. *Betweenness centrality* on the other hand, measures the incidence of a node or an edge on the shortest paths between any pair of nodes in the network. A low degree node may be connecting two groups of nodes and may have a crucial effect for the communication between these groups as in the case of a vertex that is a *cut point*. We will be investigating centralities and algorithms to calculate them in complex networks in more detail in Chapter 5.

4.6 Network Motifs

Many complex networks contain small subgraphs called *network motifs* that appear more frequently than expected in a random network. The frequency of a motif is the count of times the motif appears in a network. Higher frequencies of these patterns are possibly due to their important functions and purpose. Different complex networks may have different network motifs; however, similar network types usually have the same network motifs reflecting the underlying network generating process. In other words, these similar networks may have evolved from the same basic network types. For example, gene regulation transcriptional networks and neural networks have been observed to contain network motifs called *feed forward loops* and *bifans*. This may mean that these two networks are similar and may have common origins [30]. Finding these network motifs is a fundamental research problem in complex networks, especially in biological networks such as the PPI networks. Detecting them is closely related to the subgraph isomorphism problem which is NPC as we will see in Chapter 9.

Motifs can be identified by comparing the occurrences of these frequently appearing subgraphs in the real complex network under consideration against a random network model of the same size that basically has a similar topology. Network motifs may exist in undirected graphs and directed graphs and possible motif numbers for a given vertex number k in directed graphs is much greater than in undirected graphs. Figure 4.7.a and (b) display possible motifs in a 3-node undirected graph and two of the possible 13 configurations of motifs in a 3-node directed graph are shown in (c) and (d) where the motif in (c) is the feed forward loop.

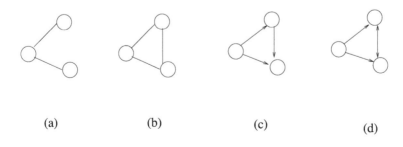

(a)   (b)   (c)   (d)

Figure 4.7: Network motifs with three nodes

4.7 Models

The models of complex networks can broadly be classified as *random networks*, *small world networks* and *scale free networks* which are described below.

4.7.1 Classical Random Networks

The basic random network model (ER-model) was proposed by Erdos and Renyi [3] based on the principle that the probability to have an edge between any pair of nodes is distributed uniformly at random. In ER-model, for a simple graph $G(V,E)$ with n vertices and m edges, there is a possible maximum of $n(n-1)/2$ edges that may exist. Each edge in this model is chosen randomly with probability $p = 2m/(n(n-1))$. The degree distribution of ER-model is binomial, approaching Poisson distribution as $n \to \infty$. The probability of a vertex with a degree k in these networks is [7]:

$$p(k) \sim e^{-k} \frac{\delta_{av}^k}{k!} \qquad (4.12)$$

The ER-model remains an early fundamental network model; however, it does not represent many real complex networks as it exhibits homogeneous degree distribution and low clustering coefficients due to being uniform. It does however exhibit a small diameter as observed in various complex networks. In order to overcome the low clustering present in ER networks, *geometric random graphs* are introduced where nodes correspond to uniformly randomly distributed points in a metric space and edges are created between pairs of nodes if the corresponding points are close enough in the metric space according to some distance norm [9]. This model has the Possion degree distribution; achieves high clustering coefficients and low average distances; however, it fails to produce power-law degree distribution as observed in many real networks.

4.7.2 Small World Networks

Many large scale complex networks have a small average distance compared to their huge sizes, enabling them to go from one vertex to another using few number of vertices. These networks are called *small-world networks*, the term originating from social network analysis which asserts that all people in the world are connected to each other with a small number of intermediate acquaintances. Sociologist Milgram concluded that people in the world are connected through a maximum of 6 people which is phrased as *six degrees of separation* [7]. Formally, the diameter of the network increases with the logarithm of the network size, that is, $d \approx \log n$ as $n \to \infty$ in small-world networks.

The ER-model exhibits small-world property as we have seen. However, many real complex networks have the local clustering property together with the small-world property which is not represented by the ER-model. Watts and Strogatz proposed a model that has a small diameter and a high local clustering at the same time [19]. In this model, the starting network is a ring topology where each node is

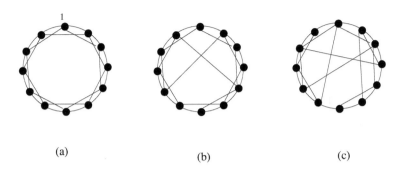

Figure 4.8: WS-network. a) The original regular network. b) Few rewiring preserves clustering and results in low average path length. c) Many rewiring approaches ER-model with low clustering and low average path length

connected to its closest $n/2$ neighbors. This type of connection would provide high local clustering as shown in real networks. In order to obtain a low average diameter in this network, rewiring of the links is provided. A link is rewired by detaching one of its ends from its originally connected node with a probability p_{rw} and connecting it to a randomly chosen vertex as shown in Figure 4.8.

As p_{rw} increases, this model approaches ER-model which means average path length becomes proportional to the network size as in the ER-model. For $p_{rw} = 0$, the regular topology is preserved, and for $p_{rw} = 1$, the ER-model is obtained. Other values of p_{rw} in WS-model result in intermediate structures between a regular and a random network exhibiting small local clustering and an average path length which is proportional to the logarithm of the network size, as observed in many real networks. However, the WS-model does not account for other features of complex networks described in the next section.

4.7.3   Scale-Free Networks

Many real-life complex networks dynamically grow and change by the addition of new nodes and edges. In order to capture this dynamic behavior, we need a model that describes the connection of the new nodes to the existing ones. This model should also provide insight to the *scale-free* property exhibited by these networks. This property we saw in Section 4.2 is characterized by power-law degree distributions as follows:

$$p(k) \sim k^{-\gamma} \text{ with } \gamma > 1, \tag{4.13}$$

meaning the probability of finding a node with a high degree diminishes as the magnitude of the degree increases. Based on the observations of the Internet and the Web, Barabasi and Albert proposed a model (BA-model) for scale-free networks with the *preferential attachment rule* [2]. Preferential attachment assumes newly added nodes to the network establish their connections preferentially with nodes of high degrees. This assumption means that nodes with initially high degrees in such a network will have continuously increased degrees as new nodes enter the network, which is also known as the *rich-get-richer* phenomenon. An algorithm to generate a network based on BA-model is shown in Alg. 4.1 where new vertices in the set V_{new} are attached to existing vertices with probabilities related to their degrees. The output graph formed using this algorithm is called *BA-network*.

Algorithm 4.1 *BA_Gen*

Input : $G(V,E)$, V_{new} : new vertices to be joined to G
while $V_{new} \neq \emptyset$ **do**
  pick $v \in V_{new}$
  $V_{new} \leftarrow V_{new} \setminus \{v\}$
  $V \leftarrow V \cup \{v\}$
  $m \leftarrow$ number of edges
  while $m \geq 1$ **do**
    attach v to $u \in V$ with probability $P(u) = \frac{\delta(u)}{\sum_{w \in V} \delta(w)}$
    $m \leftarrow m - 1$
  end while
end while

Figure 4.9 displays the operation of preferential attachment rule in an example network where black nodes are the new nodes entering the network. The scale-free networks are more robust to preserve connectivity by random edge removals, due to the large number of low-degree nodes than random networks, but are more sensitive to attacks aimed at the high-degree nodes. The highly connected nodes of scale-free complex networks presumably have more important functionality than low-degree nodes. The scale-free networks exhibit the following properties:

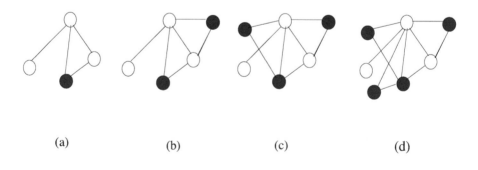

(a)   (b)   (c)   (d)

Figure 4.9: The preferential attachment rule

■ Degree distribution follows power law, with the exhibition of very few high degree nodes and many low degree nodes.

■ The average clustering coefficient of these networks is low due to the large number of low-degree nodes.

■ The average diameter is low due to the clustering of nodes around the high-degree nodes.

The BA-model describes the scale-free property observed in many real complex networks. For example, in a social network, people with many friends (high-degree nodes) have a higher probability to have new friends than people who have fewer friends. It however has the disadvantage that the new nodes entering the network should have the complete network topology information to be able to decide which nodes to connect. Variations of this model allow new node attachments based on partial knowledge of the network [4]. Certain modifications to this model involve the provision of *aging* in social networks or *capacity restriction* in technological networks. Aging provides the exponential cutoff at high degrees where nodes with high degrees stop receiving new neighbors after some time as observed in many real social networks. An example of capacity restriction in a real complex network is the airline route network where passengers are rejected by the airline after a certain capacity [7].

4.8　Chapter Notes

We have reviewed fundamental concepts in the analysis of complex networks in this chapter. We have seen that degrees of nodes, degree distribution and the clustering coefficients provide us with important insight about the structure of a network. Further network statistics and advanced aspects can be found in [7]. Centrality of nodes and edges show the importance of nodes and edges in a complex network and therefore we can distinguish nodes and edges that have more importance in the overall functioning of the complex network by calculating various centralities.

The structure of the communities or clusters which consist of nodes that are closely related to each other in a complex network have direct implications about the functioning of the network, and therefore, detection of such clusters is important in determining the regions in the network with possibly imperative functionality. Another significant structure is the network motifs which are subgraphs that repeat significantly more than other subgraphs in the whole complex network. These motifs are believed to have important functioning in the overall processing performed in many real complex networks such as the PPI networks. Different than network motif search, alignment of networks attempts to find similar regions of networks under consideration. The discovered common subgraphs should be similar based on some metric and need not be exact, as observed in many biological networks. Finding such structures may imply common origins of the organisms that the graphs represent. In

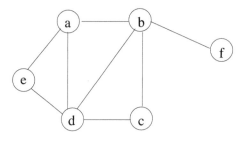

Figure 4.10: Example for Ex. 1

summary, we can conclude that the following fundamental tasks require algorithmic solutions to understand the behavior and functioning of complex networks:

■ Calculating centralities of nodes and edges

■ Detecting clusters and dense regions of complex networks

■ Discovering network motifs

■ Network alignment

Efficient algorithms to perform the above tasks in fact constitute the core of Part II in this book.

Exercises

1. What is the degree sequence of the graph in Figure 4.10? Sketch an isomorphic and a non-isomorphic graph to this graph that have the same degree sequence.

2. Plot the degree distribution of the graphs in Figure 4.11. What can be deduced about the properties of these graphs based on their degree distribution?

3. Work out the local clustering coefficients, the average clustering coefficient and the transitivity for the graphs in Figure 4.11. Comment on the relationship between the average clustering coefficients and the topologies of these graphs.

4. Write the pseudocode of an algorithm with comments, that will find the matching indices of the vertices of a graph. Extend this algorithm to find the matching indices within 2-hop neighborhoods of vertices. Find the computational complexities in both cases.

5. Work out the closeness centralities for the vertices of the graph in Figure 4.12.

6. Compare the average diameters and the average clustering coefficients of ER-model, WS-model and BA-model type of networks by giving reasons.

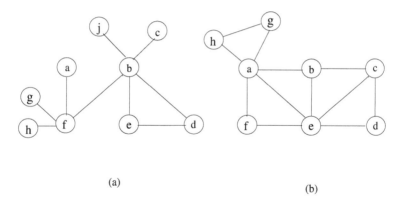

(a)   (b)

Figure 4.11: Example graphs for Ex. 2 and 3

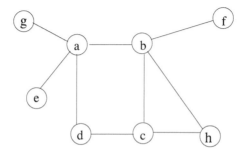

Figure 4.12: Example graph for Ex. 5

References

[1] D.A. Bader and K. Madduri, Parallel Algorithms for Evaluating Centrality Indices in Real-world Networks. In *Proceedings of the 35th International Conference on Parallel Processing (ICPP 2006)*, Columbus, OH, August 14-18, 2006.

[2] A. L. Barabasi and R. Albert. Emergence of scaling in random networks. *Science*, 286:509-512, 1999.

[3] P. Erdos and A. Renyi. On random graphs. *Publicationes Mathematicae*, 6:290-297, 1959.

[4] M. Dehmer (Ed.). *Structural Analysis of Complex Networks*, Birkenhauser, ISBN 978-0-8176-4788-9, 2011.

[5] B. H. Junker and F. Schreiber (Eds.). *Analysis of Biological Networks*. Wiley Interscience, Chapter 3, 2008.

[6] A. Li and S. Horvath. Network neighborhood analysis with the multi-node topological overlap measure. *Bioinformatics*, 23(2):222-231, 2007.

[7] S. Milgram. The small world problem. Psychology Today, 2:60-67, 1967.

[8] M. E. J. Newman. The structure and function of complex networks. *SIAM Review*, 45(2):167-256, 2003.

[9] M. Penrose. *Random Geometric Graphs (Oxford Studies in Probability)*. Oxford University Press, USA, 2003.

[10] N. Przulj. Graph theory analysis of protein-protein interactions. A chapter in *Knowledge Discovery in Proteomics*, edited by Igor Jurisica and Dennis Wigle, CRC Press, 2005.

[11] D. J. Watts and S. H. Strogatz. Collective dynamics of small-world networks. *Nature*, 393:440-442, 1998.

[12] H. Zhou. Network landscape from a Brownian particles perspective. *Physical Review E*, 67:041908, 2003.

ALGORITHMS

Chapter 5

Distance and Centrality

5.1 Introduction

We have seen in Chapter 2 that distance between two vertices u and v in a graph $G(V,E)$ is the shortest path between them. This distance, denoted by $d(u,v)$ is the number of hops of the shortest path if G is unweighted, or is the sum of the weights assigned to edges of the shortest path if it is weighted. In this chapter, we will first describe the average distance of a graph and then show an algorithm to find distances from a single node of the network to all other nodes called *single source shortest path* (SSSP) algorithm. Distances between every pair of nodes are provided by *all pairs shortest path* (APSP) algorithms and we describe an example APSP algorithm.

A fundamental problem in the analysis of complex networks is the determination of how important a certain vertex or an edge is. *Centrality* of a node in a network shows how influential that node is in the overall network. In Section 5.3, different types of centrality measures are outlined with algorithms that determine these centrality values. Modified distance finding algorithms are commonly used to evaluate centrality of vertices and edges in complex networks.

5.2 Finding Distances

We have seen in Chapter 2 that the *eccentricity* of a vertex v in a connected graph is the maximum distance from that vertex to any other which can be shown as:

$$\sigma(v) = \max_{u \in V} d(u,v) \qquad (5.1)$$

The *diameter* $(diam(G))$ of a connected graph is the maximum distance between its two vertices. It can be related to the eccentricity as follows:

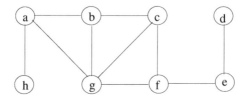

Figure 5.1: Diameter and eccentricities in a graph

$$diam(G) = \max_{v \in V} \sigma(v) \tag{5.2}$$

which states that the diameter is the maximum value of all eccentricities. The *radius* ($rad(G)$) of a graph is the minimum eccentricity among all vertices as shown below:

$$rad(G) = \min_{v \in V} \sigma(G) \tag{5.3}$$

The relation between the diameter and the radius is specified as follows:

$$rad(G) \leq diam(G) \leq 2rad(G) \tag{5.4}$$

Figure 5.1 displays a graph with diameter 5 as *h-a-g-f-e-d* is the longest path. The eccentricities of vertices $a, ..., h$ are 4, 4, 3, 5, 4, 3, 3, 5 respectively. The radius is 3 as the minimum of the eccentricities.

The diameter and radius of a complete graph K_n where $n \geq 2$ is 1; the diameter and radius of a complete bipartite graph $K_{m,n}$ is 2 where either m or n is ≥ 2 as shown in Figure 5.2.

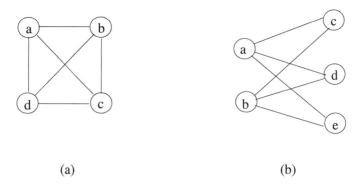

(a)   (b)

Figure 5.2: Diameters and radii of complete graphs

5.2.1 Average Distance

The average distance of a graph is another parameter that gives us useful information about the structure of the graph. We need to define the average distance of a vertex first.

Definition 6 (Average Distance of a Vertex) *The average distance $d_v(av)$ of a vertex v of a graph G is defined as the arithmetic average of the distance of v to all other vertices a follows:*

$$d_v(av) = \frac{1}{n-1} \sum_{u \in V} d(u, v) \tag{5.5}$$

The average distance of a graph shows us how easy it is to reach from one vertex to another, a low average distance meaning there are short paths between most of the vertices, and a high average distance showing it is in general difficult to reach from one vertex to another in such a graph. We can define the average distance of a graph as follows:

Definition 7 (Average Distance of a Graph) *The average distance $d_G(av)$ of a graph G is defined as the arithmetic average of the average distances of all vertices as follows:*

$$d_G(av) = \frac{1}{n} \sum_{v \in V} d_v(av) \tag{5.6}$$

Substituting Eqn. (5.5) in Eqn. (5.6) yields:

$$d_G(av) = \frac{1}{n} \sum_{v \in V} d_v(av) = \frac{1}{n} \sum_{v \in V} (\frac{1}{n-1} \sum_{u \in V} d(uv)) \tag{5.7}$$

We need efficient algorithms to compute distances between the nodes of a complex network to be able to compute average distances. The algorithm due to Dijkstra finds distances between a node and all other nodes in a computer network using the greedy approach we have described in Chapter 3 and the Floyd-Warshall algorithm that finds distances between all pairs of nodes in the network using dynamic programming is described next.

5.2.2 Dijkstra's Single Source Shortest Paths Algorithm

Dijkstra's shortest paths algorithm (*Dijkstra_SSSP*) is a generalization of the BFS algorithm for weighted graphs. The idea of this algorithm is to greedily search for the minimum distance nodes to the source and include them in the found paths. It starts by setting distances of all nodes except the source, to infinity (∞) and predecessors of them are set to undetermined ($\perp$) values. The source node s has a distance 0 to itself and its predecessor node is itself. The algorithm picks the node u with the lowest distance from unsearched nodes that are inserted in the queue S and includes u in the list of found shortest paths. For any neighbor node v of u, its distance is modified if this is longer than the the path through u to the source node s. When the distance is changed for a lower value, the predecessor node of v is changed to u to show the current effective path, as shown in Alg. 5.1 and Figure 5.3 displays a network where *Dijkstra_SSSP* algorithm is executed iteratively to find shortest paths from the source vertex a.

Algorithm 5.1 *Dijkstra_SSSP*

1: **Input** : $G(V, E)$, directed or undirected and with positive weights (l_e) on edges
2: **Output** : d_v and $pred[v]$ $\forall v \in V$ ▷ shows distance and place of a vertex in BFS tree
3: **for all** $u \in V \setminus \{s\}$ **do** ▷ initialize distances
4:     $d_u \leftarrow \infty$
5:     $pred[u] \leftarrow \perp$
6: **end for**
7: $d_s \leftarrow 0$
8: $pred[s] \leftarrow s$
9: $S \leftarrow make\_que(V)$ ▷ insert all vertices in queue S
10: **while** $S \neq \emptyset$ **do**
11:     $u \leftarrow min(S)$ ▷ find minimum distance vertex in S
12:     $S \leftarrow S \setminus \{u\}$
13:     **for all** $(u, v) \in E$ **do** ▷ update each neighbor distance of u
14:         **if** $d_v > d_u + l(u, v)$ **then**
15:             $d_v \leftarrow d_u + l(u, v)$
16:             $pred[v] \leftarrow u$
17:         **end if**
18:     **end for**
19: **end while**

Theorem 5.1
The time complexity of Dijkstra_SSSP algorithm is $O(n^2)$ for a graph of order n and size m.

Proof 7 *We have two loops where the inner loop may be executed at most n times and the outer loops is executed n times resulting in $O(n^2)$ time complexity.*

The Fibonacci heap and priority queue based implementation of this algorithm has a lower time complexity of $O(m + n \log n)$ [10].

5.2.3 Floyd-Warshall All Pairs Shortest Paths Algorithm

For graphs with negative or positive weights, the Floyd-Warshall algorithm (*FW_APSP*) can be used. This algorithm finds APSP routes in a graph using dynamic

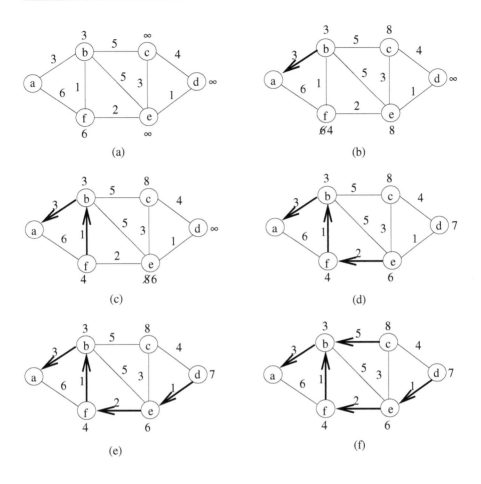

Figure 5.3: *Dijkstra_SSSP* **algorithm execution. The vertices on the shortest path to source vertex** *a* **are shown by arrows**

programming by comparing all possible paths between each pair of nodes in *G* by improving the shortest path between them at each step until the result is optimal. Starting by a single edge, it iteratively computes paths by increasing the set of intermediate nodes by adding a pivot node to the intermediate nodes.

The distance matrix $D[n,n]$ shows the current distance between the two nodes *u* and *v* and the matrix $P[n,n]$ shows the first node on the current shortest path from *u* to *v*. For every node u, v and $w \in V$, the algorithm checks whether distance through *w* will result in a shorter distance between the node *u* and *v* than the current one and if the new distance is shorter, it is stored in *D* and *P* is also modified to show the route through *w* as shown in Alg. 5.2 [6]. All of the nested loops in the algorithm are executed n times resulting in $O(n^3)$ time including initialization which also takes n^2 steps.

Algorithm 5.2 *FW_APSP*

1: $S \leftarrow \emptyset$
2: **for all** $\{u,v\} \in V$ **do** ▷ initialize
3:     **if** $u = v$ **then**
4:         $D[u,v] \leftarrow \emptyset, P[u,v] \leftarrow \perp$
5:     **else if** $(u,v) \in E$ **then**
6:         $D[u,v] \leftarrow w_{uv}, P[u,v] \leftarrow v$
7:     **else** $D[u,v] \leftarrow \infty, P[u,v] \leftarrow \perp$
8:     **end if**
9: **end for**
10: **while** $S \neq V$ **do**
11:     **pick** w from $V \setminus S$
12:     **for all** $u \in V$ **do** ▷ Execute a global w-pivot
13:         **for all** $v \in V$ **do** ▷ Execute a local w-pivot at u
14:             **if** $D[u,w] + D[w,v] < D[u,v]$ **then**
15:                 $D[u,v] \leftarrow D[u,w] + D[w,v]$
16:                 $P[u,v] \leftarrow P[u,w]$
17:             **end if**
18:         **end for**
19:     **end for**
20:     $S \leftarrow S \cup \{w\}$
21: **end while**

Example

Figure 5.4 shows a sample directed and weighted graph, and after three iterations, shortest distances between every pair of nodes are found as shown by the changes in matrix D.

$$
D = \begin{bmatrix}
0 & \infty & \infty & \infty & 2 \\
4 & 0 & 9 & \infty & \infty \\
\infty & \infty & 0 & 3 & \infty \\
\infty & 6 & \infty & 0 & \infty \\
\infty & 1 & \infty & 12 & 0
\end{bmatrix}
\rightarrow
\begin{bmatrix}
0 & 3 & \infty & 14 & 2 \\
4 & 0 & 9 & 12 & 6 \\
\infty & 9 & 0 & 3 & \infty \\
10 & 6 & 15 & 0 & \infty \\
5 & 1 & 10 & 12 & 0
\end{bmatrix}
$$

$$
\rightarrow
\begin{bmatrix}
0 & 3 & 12 & 14 & 2 \\
4 & 0 & 9 & 12 & 6 \\
13 & 9 & 0 & 3 & 10 \\
10 & 6 & 15 & 0 & 12 \\
5 & 1 & 10 & 12 & 0
\end{bmatrix}
$$

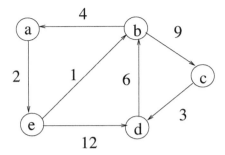

Figure 5.4: *FW_APSP* **execution example**

5.3 Centrality

We would like to find out the importance of specific nodes or edges in the network as a significantly more important node or an edge may be joining two distinct parts of the network, and may be involved in data transfers more than the others, or it may act as a source of important knowledge or ideas as in the case of a social network. The aim of assessing *centrality* of a node or an edge is to capture the importance of nodes and edges in the network to some extent. Types of centralities we will consider are the degree, closeness, stress, betweenness and eigenvalue centralities as described next.

5.3.1 Degree Centrality

A simple way of specifying how important a node in a network is related to its degree. It is expected that a higher degree node is involved in more frequent communications than lower ones as it has more neighbors. Given an undirected graph $G(V,E)$, *degree centrality* ($C_D(v)$) for a vertex v is simply its degree, $deg(v)$. For directed graphs, two parameters as *in-degree centrality* and *out-degree centrality* can be defined. This simple measure is based on the neighborhood of a node. Formally, given a graph $G(V,E)$ with n vertices and adjacency matrix $A[n,n]$, the degree centrality of a vertex i is:

$$C_D(i) = \sum_{j \in V} a_{ij} \tag{5.8}$$

where a_{ij} is the ijth entry in A. We can state that $C_D = A \times [1]$ where C_D is the vector of all degrees and $[1]$ is a vector of all 1s. Figure 5.5 displays a sample network with 6 nodes labeled $a, ..., f$.

The equation $C_D = A \times [1]$ for this network where rows and columns are ordered as $a, .., f$ can be stated as follows.

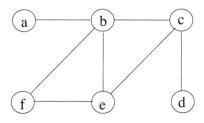

Figure 5.5: Degree centrality example

$$
\begin{bmatrix} 1 \\ 4 \\ 3 \\ 1 \\ 3 \\ 2 \end{bmatrix}
=
\begin{bmatrix}
0 & 1 & 0 & 0 & 0 & 0 \\
1 & 0 & 1 & 0 & 1 & 1 \\
0 & 1 & 0 & 1 & 1 & 0 \\
0 & 0 & 1 & 0 & 0 & 0 \\
0 & 1 & 1 & 0 & 0 & 1 \\
0 & 1 & 0 & 0 & 1 & 0
\end{bmatrix}
\times
\begin{bmatrix} 1 \\ 1 \\ 1 \\ 1 \\ 1 \\ 1 \end{bmatrix}
\tag{5.9}
$$

Given a graph $G(V,E)$ represented by an adjacency matrix A, degree centrality vector C_D can be formed by summing each row of A in n steps for a total of $\Theta(n^2)$ time. Further elaboration on degree centrality of a node can be done using the following metrics.

Definition 5.1 k-path centrality   *k-path centrality* of a node v, $(k - C_D(v))$, is the number of paths of length k or less emanating from node v [17].

When $k=1$, k-path centrality is equal to degree centrality. Edge disjoint paths do not have any common edges and vertex disjoint paths do not have any common vertices.

Definition 5.2 edge disjoint k-path centrality   *Edge-disjoint k-path centrality* of a node v is the number of edge-disjoint paths of length k or less that start or end at vertex v [3].

Ford-Fulkerson Theorem [9] states that the number of edge disjoint paths between two nodes u and v of a graph is equal to the minimum number of vertices that must be removed to disconnect u and v.

Definition 5.3 vertex disjoint k-path centrality   *Vertex-disjoint k-path centrality* of a node v is the number of vertex-disjoint paths of length k or less that start or end at vertex v [3].

Menger showed that the number of vertex disjoint paths between two nodes u and v equals the number of nodes that must be removed to disconnect u and v [14].

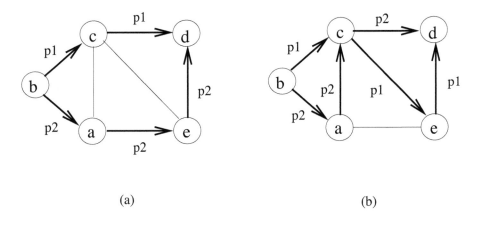

(a)   (b)

Figure 5.6: Edge and vertex disjoint paths

When $k=1$, $k - C_D(v) = C_D(v)$ and when $k = n - 1$, $k - C_D(v)$ includes all of the paths emanating from node v as n-1 is the longest path in any graph. Figure 5.6.a shows a simple network where $(b - c - d)$ and $(b - a - e - d)$ are two edge-disjoint and also vertex-disjoint paths between vertices b and d, excluding the starting and ending vertices. In (b), two edge disjoint paths between vertices b and d which are not vertex disjoint are $(b - c - e - d)$ and $(b - a - c - d)$.

A major drawback with degree centrality is that it only considers local information. For example, a node having few important neighbors may be more influential globally than a node which has many less important neighbors. In order to overcome this problem to some extent, we propose and define k-hop degree centrality $(k - hop - C_D)$ as the number of neighbors a node has in its k-hop neighborhood. For $k=1$, $k - hop - C_D$ equals C_D. We further define k-rank of a node as its position in the sorted degree list of neighbors in its k-hop neighborhood. Figure 5.7 shows a simple network with 7 vertices 1-ranks of which are shown next to them. The 1-rank values may give unexpected results for isolated nodes, however, k-rank values for $k > 1$, would provide more realistic importance of a node in a wider neighborhood. In Figure 5.7, although vertex e is isolated, it has a rank of 2 but when two hops are considered, its 2-rank value is 5.

5.3.1.1 A Distributed Algorithm for k-hop Degree Centrality

In this section, we propose a distributed algorithm to find $2 - hop - C_D$ values of the nodes of a computer network. Each node in this algorithm exchanges its 1-hop degree value with its neighbors. After this phase, sorting the values yields the 2-rank value of a node in its 2-neighborhood as shown in Alg. 5.3 and Figure 5.8. We assume each node knows the degrees of its neighbors initially, otherwise, a first step where degrees are exchanged is needed.

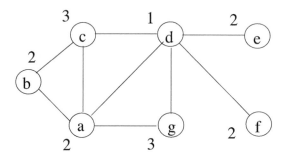

Figure 5.7: One-rank centrality example

Algorithm 5.3 *Two_rank*

1: **int** i, j ▷ i is this node, j is the sender
2: **set of int** *neighs_degs, my_degs, received* $\leftarrow \{\varnothing\}$
3: **message types** *deg*
4: **send** *deg* $(my\_degs(i))$ to $\Gamma(i)$ ▷ start
5: **while** *received* $\neq \Gamma(i)$ **do**   ▷ receive messages from all neighbors
6:     **receive** $deg(j, deg(my\_degs(j))$
7:     *received* $\leftarrow$ *received* $\cup \{j\}$
8:     *neigh_degs* $\leftarrow$ *neigh_degs* $\cup \{my\_degs(j)\}$
9: **end while**
10: **sort** *neigh_degs* in descending order   ▷ find rank
11: *my_2rank* $\leftarrow$ my rank in sorted *neigh_degs*

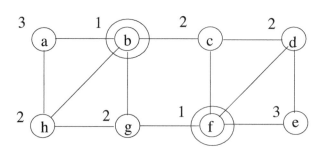

Figure 5.8: Output of *Two_rank* **algorithm**

5.3.2 Closeness Centrality

In order to provide a centrality parameter based on global knowledge rather than local, we may check the easiness of reaching all other nodes from a certain node. *Closeness centrality* of a node i in a graph $G(V, E)$ is defined as the reciprocal of the total distance from this node to all other nodes as shown below:

$$C_C(i) = \frac{1}{\sum_{j \in V} d(i, j)} \tag{5.10}$$

We can find distances from a single node to all other nodes of an undirected and unweighted graph as the number of hops by the BFS algorithm of Section 3.4. A simple modification to this algorithm to find closeness centralities of nodes would then be counting of the hops in each shortest path originating from the node s. For weighted graphs, distances between all pairs can be found by Dijkstra's shortest path algorithm.

For unweighted graphs, let us assume distances computed using the BFS algorithm in hops between each pair of vertices is stored in matrix D where d_{ij} is the number of hops of the shortest path between vertices i and j. In this case, $S = D \times [1]$ where S is the vector showing total distances from each node to all other nodes and $[1]$ is a vector of all 1s. Taking reciprocal of the elements of vector S would then yield the closeness centrality vector C_C. Figure 5.9 displays an undirected, unweighted graph with 6 vertices and BFS trees for each vertex are shown.

For this graph, inserting the values for D matrix where rows and columns are ordered as $a, .., f$ and multiplication yields the matrix S as follows:

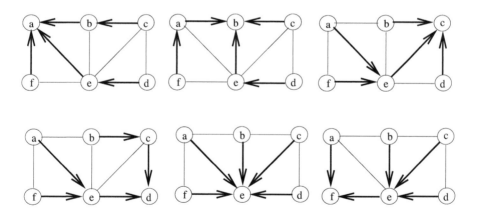

Figure 5.9: Closeness centrality calculation

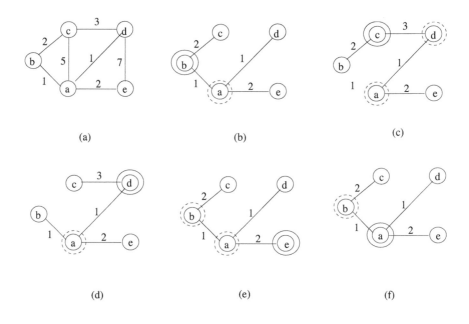

(a) (b) (c)

(d) (e) (f)

Figure 5.10: Closeness centrality in a weighted graph

$$
\begin{bmatrix} 7 \\ 7 \\ 7 \\ 8 \\ 5 \\ 8 \end{bmatrix} = \begin{bmatrix} 0 & 1 & 2 & 2 & 1 & 1 \\ 1 & 0 & 1 & 2 & 1 & 2 \\ 2 & 1 & 0 & 1 & 1 & 2 \\ 2 & 2 & 1 & 0 & 1 & 2 \\ 1 & 1 & 1 & 1 & 0 & 1 \\ 1 & 2 & 2 & 2 & 1 & 0 \end{bmatrix} \times \begin{bmatrix} 1 \\ 1 \\ 1 \\ 1 \\ 1 \\ 1 \end{bmatrix} \tag{5.11}
$$

The values of C_C for vertices $a, ..., f$ are then 1/7, 1/7, 1/7, 1/8, 1/5 and 1/8 consecutively resulting in node e with the highest closeness centrality. In Figure 5.10, weights are assigned to edges in the example network of (a) and the shortest paths starting by the double circled vertices are shown in (b),...,(f). The C_C values for the nodes $a, ..., e$ are 1/7, 1/8, 1/15, 1/9, 1/13 and we can deduce that node a is more important than others in terms of closeness centrality. We could have easily detected visually that a has the closest distance to all nodes in this small graph but in a large complex network, this would be very difficult.

5.3.3 Stress Centrality

Total number of all pairs shortest paths that pass through a vertex v is called its *stress centrality* defined as follows:

$$
C_S(v) = \sum_{s \neq t \neq v} \sigma_{st(v)} \tag{5.12}
$$

where $\sigma_{st}(v)$ is the number of shortest paths between vertices s and t. The number of shortest paths through a vertex v will show an estimate of the stress that a node in a network bears, assuming all communication will be carried along the shortest paths. APSP algorithms which are modified to count the number of shortest paths through vertices can be used to calculate stress centralities of nodes.

5.3.4 Betweenness Centrality

The *vertex betweenness centrality* or *betweenness centrality* (C_B) is more frequently used than other metrics for assessing the influence of a node and is based on the idea that if the paths for shortest paths between nodes of a network pass through some vertices more often than others, then these vertices are significantly more important than others for communication purposes. Formally, C_B for a node $v \in V$ of a graph $G(V, E)$ is given by [11]:

$$C_B(v) = \sum_{s \neq t \neq v} \frac{\sigma_{st}(v)}{\sigma_{st}} \tag{5.13}$$

where σ_{st} is the total number of shortest paths between vertices s and t, and $\sigma_{st}(v)$ is the total number of shortest paths between vertices s and t that pass through vertex v. Betweenness centrality may be regarded as the normalized stress centrality. Checking Figure 5.10 again, the intermediate vertices on the shortest paths are shown by dotted circles and therefore only the vertices a, b and d have nonzero betweenness values. There are $n(n-1)$ shortest paths in a network with n nodes, counting in both directions. We have a total of 20 shortest paths in this graph, and the C_B values for vertices $a, .., e$ are then $0.7, 0.2, 0, 0.2, 0$ consecutively and we can deduce that vertex a is more important than others in this graph as it has the highest value which can also be detected visually as it is an intermediate vertex of several shortest paths.

A simple procedure to find C_B for a node v is then to find all shortest paths in the graph G and count the number of paths that pass through v by excluding the paths that start or end at v. For an unweighted graph G, we could simply run a modified BFS algorithm for each node in the graph and find the shortest paths between each vertex pair (u, v) as in the BFS algorithm but if there are two or more paths of the same length between a vertex u and the root vertex s, then u is assigned as a child of all its neighbor vertices that have the same distance to s. In other words, a vertex u may have more than one parent if all of these parents are equidistant to s. We will call the resulting directed graph the *BFS structure* as it is not a tree anymore due to possible cycles. As the time complexity of BFS is $O(n+m)$, total time is $O(n(n+m))$ for all nodes. The pair dependencies can then be calculated as follows:

$$\delta_{st} = \frac{\sigma_{st}(v)}{\sigma_{st}} \tag{5.14}$$

The final step of this procedure is the summation of all the pair dependencies. For weighted graphs, Dijkstra's SSSP algorithm or Floyd-Warshall APSP algorithm may be modified to compute pair dependencies as described above which yields $\Theta(n^3)$ time steps.

5.3.4.1 Newman's Algorithm

Newman proposed an algorithm based on the BFS algorithm to find C_B values of the nodes of a graph with significantly less time complexity than the described procedures. This algorithm consists of the following steps [16]:

1. **forall** $v \in V$

2. $b_v \leftarrow 1$

3.   **find** BFS paths from v to all other nodes

4. **end**

5. **forall** $s \in V$

6.   Starting from the farthest nodes, move from u towards s along shortest paths using vertex v

7. $b_v \leftarrow b_v + b_u$

8.   If v has more than one predecessor, then b_v is divided equally between them.

9. **end**

10. $C_B(v) = \sum_{i=1}^{n} b_v$

The algorithm should exclude the starting and ending nodes. The reason to divide b_v between the parents of a node is to compute the effect of multiple shortest paths. In other words, if there are k parent nodes from a node u to the source node s, then these nodes share the amount of data transfer and communication between u and s equally. Figure 5.11 shows a sample undirected network and the BFS structures for all nodes $a, ..., f$ are shown where the source node is shown in double circles.

As can be seen, some nodes in each BFS structure have two parents as both of these parents are on the shortest paths from these nodes to the source node. To implement the vertex betweenness algorithm in this example, we first assign b_v values of unity to each node. We then start from each leaf node and apply the rules in the algorithm. For example, starting from node d towards the source node a in Figure 5.11.a, node d contributes 0.5 to both of its parents raising their betweenness values to 1.5. Furthermore, node c also contributes 1 to b as it is its single parent raising the value of node b to 2.5. The start and end points of a shortest path are not given weights and proceeding similarly for other nodes, the vertex betweenness values at each iteration found are shown next to each node, resulting in the final vertex betweenness values for nodes $a, ..., f$ as 5, 14, 0, 0, 10, and 0 consecutively. Node b is the most influential node in this network which can be seen visually as it is on several shortest paths. If

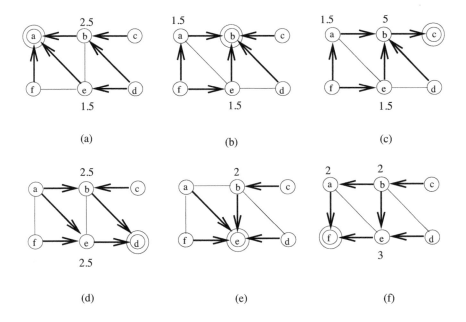

Figure 5.11: Betweenness centrality algorithm example

we subtract the number of times that a node is incident on any shortest path in the BFS graphs from its betweenness value, we find the number of shortest paths that pass through that node. For example, node b has a betweenness value of 14 and in five graphs of Figure 5.11, it is an intermediate node for 5 shortest paths. Subtracting 5 from 14 gives the total number of shortest paths that pass through node b which is 9. We can obtain the same result if b_v value for any intermediate node is set to 0. This way of calculating the values in fact removes the initial values of 1 from any intermediate vertices. The time complexity of this algorithms is $O(n(n+m))$ as in the BFS algorithm for APSPs in an undirected network.

5.3.4.2 Brandes' Algorithm

Brandes proposed an algorithm to compute betweenness centralities of the vertices of a graph in $O(mn+n^2\log n)$ time for weighted graphs, and $O(mn)$ time for unweighted graphs [4]. The pair dependency between nodes s,t in this algorithm is defined as δ_{st} as before. Brandes however, introduced the dependence of a vertex s on v as:

$$\delta_{s*}(v) = \sum_{t \in V} \delta_{st}(v) \qquad (5.15)$$

Therefore, the betweenness centrality for each vertex is the sum of all of the dependency values from all vertices as:

$$C_B(v) = \sum_{s \in V} \delta_{s*}(v) \qquad (5.16)$$

In the first step of the algorithm, all of the shortest paths from each vertex to all other vertices and predecessor vertices on these shortest paths are found. In the second step of the algorithm, for every vertex $s \in V$, the dependencies $\delta(v)$ for all $v \in V$ are computed using the shortest paths trees data and predecessor sets along these paths. In the last step, the sum of all dependency values are computed to give betweenness centrality values of vertices.

The *edge betweenness* value of an edge e in a graph G is defined as the number of shortest paths that pass through e [16]. If there is more than one shortest path between a pair of vertices, each path is assumed to have equal weight such that the total weight of all the paths is unity. Edges connecting two clusters A and B in a graph will have high edge betweenness values as all of the shortest paths from any vertex u in A to any vertex v in B will pass through these edges. This will be more evident in the case of a bridge, removal of which which will disconnect the graph. We will investigate algorithms that find edge betweenness values for edges and the use of this parameter for community detection in social networks in Chapter 11.

5.3.5 Eigenvalue Centrality

A node of a network may be considered important if it has important neighbors. As a further attempt to asses the importance of a node based on the importance of its neighbors, we may assign variable x_i as the importance of node i called its *score*. Given a graph $G(V,E)$ with an adjacency matrix A, the score of a node i can be defined as proportional to the sum of all its neighbors' scores as follows:

$$x_i = \frac{1}{\lambda} \sum_{j \in N(i)} x_j = \frac{1}{\lambda} \sum_{j \in V} a_{ij} x_j \tag{5.17}$$

where $N(i)$ is the set of neighbors of i and λ is a constant. The score of a node can simply be its degree. In this case, we are in fact adding the degrees of the neighbors of a node i to find its new score in Eqn. (5.17). In vector notation,

$$\vec{x} = \frac{1}{\lambda} A \vec{x} \tag{5.18}$$

which is the eigenvalue equation $A\vec{x} = \lambda \vec{x}$. There will be n different values of the eigenvector, however, by Perron-Frobenius theorem which asserts that a real square matrix with positive entries has a unique largest real eigenvalue and that the corresponding eigenvector has strictly positive components [16]; only the greatest eigenvalue provides the desired centrality value which is called *the eigenvalue centrality*. In order to find the eigenvalue centrality of a graph G, we need to find the eigenvalues of the adjacency matrix A. We then select the the largest eigenvalue and find the associated eigenvector for this value. This eigenvector contains the eigenvalue centralities of all nodes. Eigenvalue centrality has been successfully used in Google's page rank algorithm [26].

5.4   Chapter Notes

Centrality is a fundamental parameter to analyze complex networks and it is especially used in social networks to find the importance of a node [3, 17]. Shortest distances between the nodes in a graph are needed to be able to calculate the centrality values in most cases. If the graph is unweighted, the BFS algorithm we have seen in Chapter 3 can be used. For weighted graphs, we can use Dijkstra's SSSP algorithm and Floyd-Warshall algorithm for APSP routes as described in Section 5.2. Bellman-Ford algorithm which uses dynamic programming also computes SSSP routes from a single source in $O(nm)$ time [5].

We examined degree, closeness, stress, vertex betweenness and eigenvalue centralities and the algorithms to find them in this chapter. Betweenness centralities and the eigenvalue centrality provide more realistic importance values of vertices and are implemented more often than others in the analysis of complex networks. We have also proposed and defined k-hop centrality and related k-ranks of nodes and described a distributed algorithm to compute this centrality in computer networks.

There are other measures of centrality such as Katz centrality [13] which considers all paths between nodes instead of the shortest paths only and subgraph centrality that was defined by Estrada [7] as a measure of protein folding. Efficient methods to calculate centralities are still active research areas in the study of complex networks. Although there are many sequential algorithms to compute various centralities, design and implementation of parallel and distributed algorithms for this problem is a relatively less studied area. The algorithms provided in [3] result from one of the few studies that describe parallel computation of various centralities. The scale of the problem being large, in fact necessitates the employment of parallel algorithms and for the case of computer networks such as the Internet or ad hoc wireless networks, using distributed algorithms to compute centralities is inevitable. We believe there is much work to be done in this topic, namely, the design of parallel and distributed algorithms to compute centralities.

Exercises

1. Work out the shortest paths from each node to all other nodes in the weighted graph of Figure 5.12 using Dijkstra's shortest path algorithm.

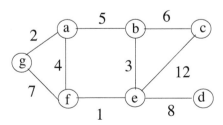

Figure 5.12: Example graph for Ex. 1 and 2

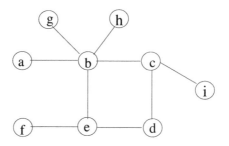

Figure 5.13: Example graph for Ex. 3

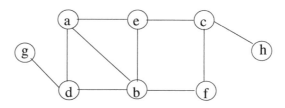

Figure 5.14: Example graph for Ex. 4

2. Work out the shortest paths from each node to all other nodes in the weighted graph of Figure 5.12 using Floyd-Warshall APSP algorithm.

3. Find the closeness centralities for all nodes in the unweighted graph of Figure 5.13 using the BFS algorithm.

4. Find the betweenness centrality values for each node in the graph of Figure 5.14 using the definition and also by the Newman algorithm and show that these two methods yield the same results.

5. For the graph of Figure 5.15, investigate and comment on the effect of multiplying the adjacency matrix A of the graph with its degree vector and attribut-

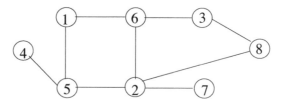

Figure 5.15: Example graph for Ex. 5

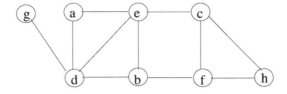

Figure 5.16: Example graph for Ex. 6

ing these values to the nodes. If the new vector formed is multiplied by A, what is achieved?

6. Find the 2-rank values for the nodes in the example graph of Figure 5.16. Show also how the distributed two-rank algorithm can be modified to find k-ranks of nodes.

References

[1] D.A. Bader and K. Madduri. Parallel algorithms for evaluating centrality indices in real-world networks. In t*he 35th International Conference on Parallel Processing (ICPP 2006)*, Columbus, OH, 2006.

[2] R. Bellman. On a routing problem. *Quarterly of Applied Mathematics*, 16:87-90, 1958.

[3] S.P. Borgatti and M.G. Everett. A Graph-theoretic perspective on centrality. *Social Networks*, 28:466-484, 2006.

[4] U. Brandes. A faster algorithm for betweenness centrality. *J. Math. Sociol.*, 25:163-177, 2001.

[5] E.W. Dijkstra. A note on two problems in connexion with graphs. *Numerische Mathematik*, 1:269-271, 1959.

[6] K. Erciyes. *Distributed Graph Algorithms for Computer Networks*. Computer Communications and Networks Series, Chapter 7, Springer, ISBN 978-1-4471-5172-2, 2013.

[7] E. Estrada. Characterization of 3D molecular structure. Chem. Phys. Lett. (319):713-718, 2000.

[8] T.H. Cormen, , C.E. Leiserson, R.L. Rivest and C. Stein. *Introduction to Algorithms* (2nd ed.). MIT Press and McGraw-Hill. ISBN 0-262-03141-8., Section 26.2, 558-565, 2001.

[9] L.R. Ford and D.R. Fulkerson. Maximal flow through a network. *Canadian Journal of Mathematics*, (8):399-404, 1956.

[10] M. L. Fredman and R.E. Tarjan. Fibonacci heaps and their uses in improved network optimization algorithms. Journal of the Association for Computing Machinery, 34(3):596-615, 1987.

[11] L. Freeman. A set of measures of centrality based on betweenness. *Sociometry*, 40(1):35-41, 1977.

[12] P. Holme, M. Huss and H. Jeong. Subnetwork hierarchies of biochemical pathways. *Bioinformatics* 19(4):532-538, 2003.

[13] L. Katz. A new status index derived from sociometric index. *Psychometrika*, (18)1:39-43, 1953.

[14] K. Menger. Zur Allgemeinen Kurventheorie. *Fund. Math.* 10, 96-115, 1927.

[15] M.E.J. Newman. Scientific collaboration networks. II. Shortest paths, weighted networks, and centrality. *Physical Review E*, 64(1):016132, 2001.

[16] O. Perron. Zur Theorie der Matrices. Mathematische Annalen 64(2):248-263, 1907.

[17] D.S. Sade. Sociometrics of macaca mulatta III: N-path centrality in grooming networks. *Social Networks* 11:273-292, 1989.

[18] http://www.ams.org/samplings/feature-column/fcarc-pagerank

[19] D.J. Watts and S.H. Strogatz, Collective dynamics of 'small-world' networks. *Nature*, 393:440-442, 1998.

Chapter 6

Special Subgraphs

6.1 Introduction

Discovering subgraph structures in a complex network may provide us with important information about the functionality of the network. We have briefly discussed in Chapter 4 that if these structures are frequent in the network, they are called network motifs and may indicate fundamental structures with basic functions in that network.

Our aim in this chapter is to discover and construct subgraphs in a complex network that may be attributed to some special function about the structure of that network. We have already described some of these structures in relation to the complexity of the algorithms; here, our aim is to investigate efficient approximate and heuristic algorithms for the solutions to these problems. We start with the *independent set* which consists of a subset of the vertices of a graph such that no member of this set is a neighbor to another vertex in this set. The second special group of vertices in a graph called *a dominating set* consists of a subset of vertices of a graph G such that each vertex of G is either in this set or a neighbor of a vertex in this set. We will see that these two concepts are related. We then investigate the *matching* in a graph which consists of a subset of edges of the graph where any vertex incident to an edge of matching is not a neighbor to any other vertex incident to another edge of the matching. Finally, we look into the vertex cover that is a subset of vertices such that each edge of the graph is incident to at least one vertex in this set. For each problem described, efficient sequential algorithms are provided and for the case of computer networks, we show the implementation of a sample distributed algorithm to construct a vertex cover. These special subgraphs are mainly used to find clusters and also for various other applications in complex networks.

6.2   Maximal Independent Sets

Constructing independent sets in graphs provides us with basic structures in complex networks that can be used for more complex operations such as clustering. Given a graph $G(V,E)$, an independent set can be formally defined as follows.

Definition 6.1 independent set   An *independent set* (*IS*) of a graph $G(V,E)$ is a vertex set $IS \in V$ such that for any vertices $u \in IS$ and $v \in IS$, $(u,v) \notin E$, that is, there is not an edge joining two elements of the set IS.

The size of an independent set is the number of vertices it contains. Finding an IS of minimal size is trivial as even a single node of a graph would be an IS. Our main concern is to find an IS of maximal order, that is, the IS cannot be enlarged any further by the addition of any other vertices. Such an IS is called a *maximal independent set* (MIS) of the graph G. The *maximum independent set* (MaxIS) is the independent set with the largest size among all possible independent sets of a given graph G and its size is shown by $\alpha(G)$. Here, we mean the number of vertices by the size of the IS. Finding the maximum independent set of a graph is an NP-hard optimization problem and deciding whether a graph has a MIS of size k is an NP-complete problem [2].

We have also seen in Chapter 3 that a set is independent if and only if it is a clique in the complement of the graph and also a set is independent if and only if its complement is a vertex cover. The sum of $\alpha(G)$ and the size of minimum vertex cover ($\beta(G)$) are the number of vertices in the graph. Figure 6.1 displays IS examples on the same graph.

As an attempt to find an algorithm that finds the IS of a given graph, we may include an arbitrary vertex u in the MIS, but whenever we include this vertex, we need to delete it along with all of its neighbors from the graph so that they cannot be included in the MIS in the further iterations of the algorithm, as shown in Alg. 6.1. The algorithm stops when there are no more vertices that can be selected.

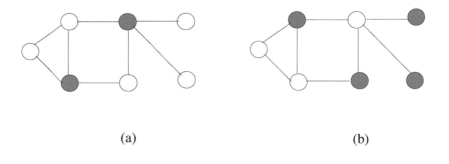

(a)   (b)

Figure 6.1: Independent set examples. a) An independent set of size 2 which is also maximal. b) A maximum independent set of size 4 in the same graph

---

Algorithm 6.1 *MIS_Alg*

1: **Input** : $G(V,E)$
2: $V' \leftarrow V, MIS \leftarrow \emptyset$
3: **while** $V' \neq \emptyset$ **do**
4:     **select** an arbitrary vertex $u \in V'$
5:     $V' \leftarrow V' \setminus \{u \cup \Gamma(u)\}$
6:     $MIS \leftarrow MIS \cup \{u\}$
7: **end while**

The two different executions of this algorithm in the same graph with 8 vertices is shown in Figure 6.2 where the sequence of iterations and the nodes deleted from the graphs are shown inside the dashed regions. The MIS elements selected are the same in both graphs, but the random selection sequence is different, resulting in different sequences of neighbor node and edge removals.

Theorem 6.1
[2] Time complexity of MIS_Alg is $O(n)$.

Proof 8 *A linear network is shown in Figure 6.3 with n nodes, black nodes are the MIS nodes, and each iteration selects nodes to be in MIS from left to right in sequence with the removed vertices and edges from the graph are shown by dashed ellipses. There will be at most $\lceil n/2 \rceil$ selected nodes for this network resulting in a total time of $O(n)$ which is in fact the worst case for this algorithm.*

6.3   Dominating Sets

A *dominating set* of a graph is a subset of its vertices such that any vertex of the graph is either in this set or is a neighbor of at least one vertex in this set. Dominating sets have many real-world applications such as clustering and routing in ad hoc wireless networks. A formal definition of a dominating set is as follows [2].

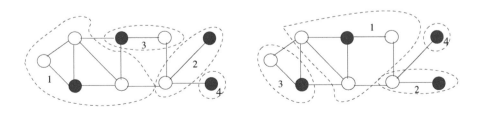

Figure 6.2: *MIS_Alg* **execution example**

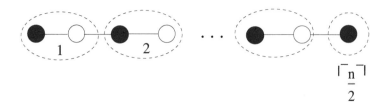

Figure 6.3: Linear network independent set example

Definition 6.2 Given a graph $G(V,E)$, a *Dominating Set (DS)* is the set of vertices $V' \in V$ such that any vertex $v \in V$ is either in V' or adjacent to a vertex in V'.

A dominating set of a graph $G(V,E)$ is *minimum (MinDS)* if it has the smallest size among all possible dominating sets of G. A dominating set is *minimal (MDS)* if the vertices in this set are not a subset of vertices in any other dominating sets of G. Finding a minimum size dominating set is NP-hard [6].

If there is a path between each pair of vertices in a dominating set, it is called a *connected dominating set* (CDS). The *minimum connected dominating set (MinCDS)* of a graph G is a minimum size dominating set of G which is connected and finding MinCDS is NP-hard. The *minimal connected dominating set (MCDS)* is a CDS that is not included in any other connected dominating sets of G. Based on the definition of a DS, Every MIS is a dominating set, however, every dominating set is not a MIS since some vertices of a DS may be neighbors. A CDS of a graph G can be constructed by first forming an MIS of G and then using intermediate vertices to connect the vertices in the MIS to form a CDS, which are included in the CDS together with the MIS vertices. Approximation algorithms are frequently used to converge to MinCDS as finding a connected minimum dominating set is an NP-Hard problem. Figure 6.4 shows a minimal dominating set and a minimum dominating set.

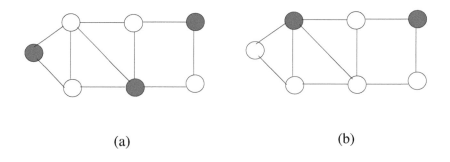

(a)   (b)

Figure 6.4: Dominating set examples. a) A minimal dominating set. b) A minimum dominating set in the same graph

In order to design algorithms to find MDS and MCDS, we will be using a simple coloring scheme to identify the status of the nodes where black nodes are the nodes in the dominating set; gray nodes are the dominated nodes each of which has at least one black neighbor and white nodes are neither dominators nor dominated. We define the *span* of a node as the number of white neighbors it has, including itself. Clearly, any algorithm that finds a dominating set of a graph should continue until there are no white nodes left.

6.3.1   A Greedy MDS Algorithm

As a first attempt, we will design a greedy algorithm that always selects a gray or white vertex with the highest span; colors it black and colors all of its white neighbors gray. The pseudocode for this algorithm called *MDS_Alg* is shown in Alg. 6.2.

Algorithm 6.2 *MDS_Alg*

1: **Input** : An undirected unweighted graph $G(V, E)$
2: **for all** $u \in V$ **do**   ▷ all nodes are white initially
3:     $color[u] \leftarrow white$
4: **end for**
5: $V' \leftarrow V, MDS \leftarrow \emptyset$
6: **while** $\exists w \in V' : color[w] = white$ **do**
7:     **select** $v \in V'$ with the highest span
8:     $color[v] \leftarrow black$
9:     $V' \leftarrow V' \setminus \{v\}$
10:     **for all** $u \in (V' \cap \Gamma(v))$ **do**
11:         **if** $color[u] = white$ **then** $color[u] \leftarrow gray$
12:     **end for**
13: **end while**

An example execution of *MDS_Alg*1 is shown in Figure 6.5 where the node with the highest span is colored black at each iteration and after two iterations, there are no white nodes left and the algorithm stops.

MDS_Alg provides a MDS as it has to continue until there are white nodes and removal of a black (MDS) node from the MDS will result in at least one of its neighbors becoming white again. The time complexity of this algorithm is $O(n)$ since we may need to have n iterations as in the case of a linear network. The approximation ratio of this algorithm is $\ln \Delta$ as shown in [9]. *MDS_Alg* may be modified so that only gray nodes with the highest scan are selected to result in a connected MDS (See Ex. 3). In certain graphs, this algorithm may yield a dominating set which may be far from optimal. For example, in a graph where two vertices u and v are connected by n different paths with one intermediate vertex in each path, this algorithm may start by coloring u or v black and then attempt to color all of the intermediate vertices

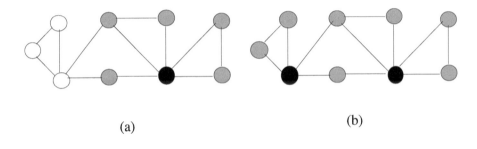

(a)   (b)

Figure 6.5: Execution of *MDS_Alg*1 **in two iterations**

black, resulting in $n-1$ steps where coloring of u and v black suffices to form a DS
in two steps. Guha-Khuller algorithms provide improvements to *MDS_Alg* so that
such cases are handled more efficiently as described in the next sections.

6.3.2   Guha-Khuller First MCDS Algorithm

In the first algorithm proposed by Guha and Khuller called *MCDS_Alg*1, we start
by coloring the highest degree vertex black and all of its neighbors gray as before.
However, we scan all the gray nodes and their white neighbors and check the spans
of gray nodes and the pair consisting of gray node and its white neighbor. We then
color either a gray node black or the pair consisting of the gray node and its white
neighbor black, depending on whichever has the highest span. We continue until all
of the nodes are colored black or gray. A sample execution of this algorithm is shown
in Figure 6.6. *MCDS_Alg*1 provides a connected MDS as we always choose a gray
node or a gray-white node pair to color black and the selected gray node will be next
to at least one black node. *MCDS_Alg*1 has an approximation ratio of $2(1 + H(\Delta))$
to the optimum CDS where H is the harmonic function as shown in [5].

6.3.3   Guha-Khuller Second MCDS Algorithm

In the second algorithm due to Guha and Khuller which we will name *MCDS_Alg*2,
a *piece* is defined as either a connected black component or a white node. The aim of
this algorithm is to always select a gray or a white node that causes the the greatest
reduction in the number of pieces in the graph. This selected node is colored black
and its white neighbors are colored gray. At the end of the first phase of the algorithm,
each node is colored black or gray, however, the resulting dominating set may not be
connected. A Steiner tree algorithm is used in the second stage to connect black
nodes to form a MCDS [5].

The execution of *MCDS_Alg*2 is shown in Figure 6.7 in a graph with 10 nodes.
The highest degree node is first colored black and all of its neighbors are colored
gray as shown in (a). In the second iteration of the algorithm shown in (b), the
node coloring of which results in lowest number of pieces is colored black with its

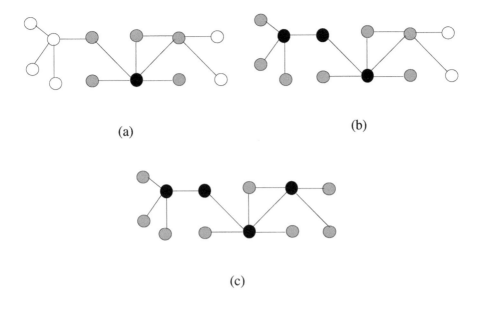

(a) (b)

(c)

Figure 6.6: Execution of *MCDS_Alg*1 **in three iterations**

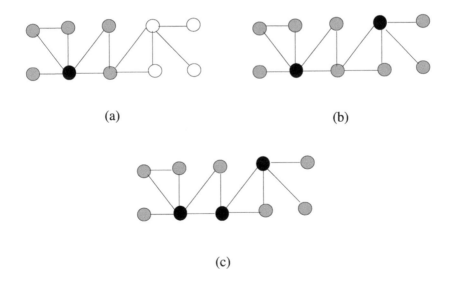

(a) (b)

(c)

Figure 6.7: Execution of *MCDS_Alg*2 **in three iterations**

neighbors colored gray, and in fact this completes the construction of MDS. In the second phase shown in (c), the intermediate gray node between the two black nodes is colored black resulting in a MCDS. *MCDS_Alg2* provides a CDS with an approximation ratio of $3 + \ln \Delta$ [5].

6.4 Matching

An independent set of edges of a graph $G(V,E)$ where the vertices incident to different edges are not neighbors is called a *matching* (M) . Matching has many applications, especially for routing and channel management in computer networks. Matching is formally defined as follows.

Definition 6.3 matching   Given an undirected graph $G(V,E)$, a matching ($M \in E$) consists of edges such that no vertex in V is incident to more than one edge in M.

A *maximal matching* (MM) of a graph $G(V,E)$ is a matching of G which cannot be enlarged by the addition of another edge. A *maximum matching* (MaxM) of G is the matching of G with the maximum size among all possible matchings of G. The *maximum weighted matching* of a weighted graph G is a matching of G where any other weighted matching of G which has a total weight larger than MaxWM does not exist. Figure 6.8 displays matching examples of a sample graph. There can be more than one MM and MaxM of a graph with the same size.

In the *bipartite matching* of a bipartite graph $G(V_1,V_2)$, we seek the matching of edges between the vertex sets V_1 and V_2 such that any edge of matching does not have any common endpoints. *Maximal bipartite matching* aims to find a maximal matching in a bipartite graph and in *maximal weighted bipartite matching* , we look for a set of matching edges that have the maximal weight.

In *perfect matching*, *(1-factor matching)* each vertex of the graph is incident to exactly one edge of the matching. An *alternating path* of a matching M is defined as a path in which the edges belong alternatively to the matching and not to the matching.

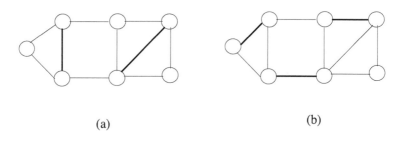

(a)   (b)

Figure 6.8: Matching examples. a) A maximal matching of a graph. b) A maximum matching of the same graph

An augmenting path is an alternating path that starts from and ends on unmatched vertices. A matching is maximum if and only if it does not have any augmenting paths [3]. Maximal unweighted or weighted matching of a graph can be performed in polynomial time using an algorithm as described by Edmonds [3].

6.4.1 A Maximal Unweighted Matching Algorithm

In order to find a maximal matching of an undirected unweighted graph $G(V,E)$, we can design an algorithm that randomly selects an edge (u,v), includes (u,v) in the maximal matching and removes edges incident to u and v from G and this operation continues until there are no edges left. The pseudocode for this algorithm called *MM_Alg* is shown in Alg. 6.3 [2].

Algorithm 6.3 *MM_Alg*

1: **Input** : undirected graph $G(V,E)$
2: $E' \leftarrow E, MM \leftarrow \emptyset$
3: **while** $E' \neq \emptyset$ **do**
4:      **select** any $(u,v) \in E'$
5:      $MM \leftarrow MM \cup \{(u,v)\}$
6:      **remove** all edges incident to u and v from E'
7: **end while**

In Figure 6.9, the operation of *MM_Alg* is shown in a sample graph with the removed edges shown as marked at each iteration. As there will be at least a single edge removal at each step, the time complexity of *MM_Alg* is $O(m)$ steps.

6.4.2 A Maximal Weighted Matching Algorithm

Given a weighted graph $G(V,E,w)$, a weighted matching algorithm should consider the weights attached to the edges of the graph.

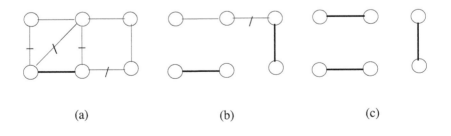

      (a)   (b)   (c)

Figure 6.9: The maximal unweighted matching algorithm example

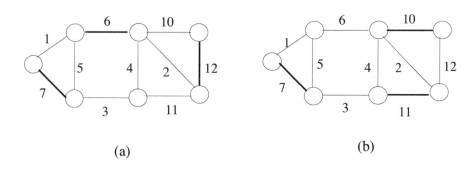

(a)   (b)

Figure 6.10: The maximal and maximum weighted matching algorithm example

We can design a greedy algorithm called *Seq_MWM* which selects the heaviest edge (u,v) from the graph at each iteration this time, instead of a random choice and removes (u,v) and all edges incident to vertices u and v from the graph. It continues until there are no edges left as in *MM_Alg*. It can be shown that this algorithm has a time complexity $O(m \log n)$ [1] with an approximation ratio 2. A different version of this algorithm was proposed in [7] where the heaviest edge is searched locally rather than globally and it was shown that this algorithm achieves an approximation ratio of 2 in $O(m)$ time.

The operation of *MWM_Alg* is shown in Figure 6.10.a where edges with weights 12, 7 and 6 are selected in sequence resulting in a total weight of 25 for this matching. However, the maximum matching as shown in (b) has edges with weights 11, 10 and 7 for a total weight of 28. The approximation ratio of *MWM_Alg* for this example is 25/28.

6.5 Vertex Cover

As an example of approximation algorithms, we have investigated the vertex cover problem in Section 3.9 in Chapter 3. A *vertex cover* of a graph G was defined as a subset of the vertices of G where each edge in the graph has at least one endpoint in this set. Vertex cover finds many applications such as facility allocation where we need to locate k resources in n places such that there is a connection between every location and every resource. The *size* of a vertex cover is the number of vertices that the cover has. Vertex cover of a graph can be defined formally as follows [2].

Definition 6.4 vertex cover   Given a graph $G(V,E)$, a *vertex cover* $VC \in V$ is the set of vertices such that for any edge $(u,v) \in E$, either $u \in VC$; $v \in VC$ or both u and v are in VC.

Finding a vertex cover of a graph is equivalent to finding an independent set of the graph such that given a graph $G(V,E)$, $IS \in V$ is an independent set if and only if

$V - IS$ is a vertex cover. For any matching M and any vertex cover VC of G, $|M| \leq |VC|$ as each edge can only cover one edge of M. In *bipartite matching* of a bipartite graph $G(V_1, V_2)$, we aim to cover all edges of G as the vertex cover of a simple graph [2].

Finding a vertex cover of maximum size is irrelevant as all of the vertices of the graph constitute a vertex cover. A *minimal vertex cover* (MVC) is a vertex cover such that removal of a vertex from MVC results in a cover that leaves at least one edge not covered. A *minimum vertex cover* (MinVC) of a graph G is the set of vertices of G that is a vertex cover with the minimum size among all possible vertex covers.

As weighted edges in weighted graphs, vertices can be assigned weights to indicate an attribute they possess in the network. In this case, we may be interested to find a vertex cover of a small sum of weights rather than a small size. A *minimal weighted vertex cover (MWVC)* of a vertex weighted graph $G(V, E, w)$ where $V : w \rightarrow \mathbb{R}$ is its vertex cover with a minimal sum of weights, that is, a vertex cover where removal of a vertex from this set results in a vertex cover which is not minimal. The minimum weighted vertex cover (MinWVC) of a weighted graph G is a set of vertices that is a vertex cover which results in a minimum total sum of vertex weights among all possible vertex covers.

Another aspect of the vertex cover is whether it is connected, that is, there is a path between each pair of vertices in the cover or not. The minimal connected vertex cover (MinCVC) is a minimal vertex cover where the graph induced by this cover in G is connected. The minimum connected vertex cover (MinCVC) is a minimum vertex cover where the graph induced by this cover in G is connected. Figure 6.11 displays MinVC and MinCVC examples. A MinVC and a MinWVC is depicted in Figure 6.12 where weights of vertices are shown next to them.

MinVC computation is an NP-hard problem [6] and there are various approximation algorithms that compute MVC, MWVC and MCVC such as the one [8] which uses depth-first search and semi-definite relaxation is used in [6]. We have already seen an approximate minimal vertex cover algorithm and a greedy algorithm which does not always approximate in Section 3.9. We will now investigate connected and

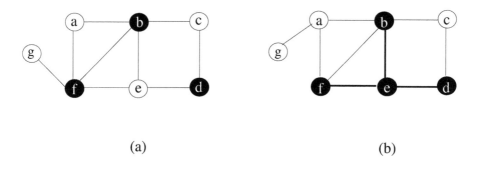

(a)   (b)

Figure 6.11: Vertex covers in the same graph. a) A minimum vertex cover of size 3. b) A minimum connected vertex cover of size 4

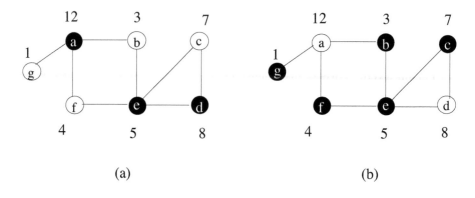

Figure 6.12: a) A minimum unweighted. b) A minimum weighted vertex cover of the same graph where weights are shown next to vertices

weighted vertex cover algorithms and conclude by describing a distributed vertex cover algorithm that can be used by the nodes of a computer network.

6.5.1 A Minimal Connected Vertex Cover Algorithm

Our aim now is to extend the approximate vertex cover algorithm of Section 3.9 to find minimal *connected* set of vertices to form the cover. We first select an arbitrary vertex $v \in V$ of the graph $G(V,E)$ and include it in the cover and remove all edges incident to this vertex from G. We then iterate as long as there are uncovered edges, by randomly selecting any neighbor of vertices in the cover that has uncovered edges. Selecting neighbors of the vertices in the cover ensures the result is a connected vertex cover. Alg. 6.4 shows the pseudocode for this algorithm called *MCVC_Alg*.

Algorithm 6.4 *MCVC_Alg*

1: **Input**: Undirected, unweighted graph $G(V,E)$
2: $S \leftarrow E$
3: $MCVC \leftarrow \emptyset$
4: **while** $S \neq \emptyset$ **do**
5:     **select** any $(u,v) \in E$ such that $v \in MCVC$ and u has uncovered edges
6:     $MCVC \leftarrow MCVC \cup \{u\}$
7:     $S \leftarrow S \setminus \{(u,v)\}$
8:     **delete** all edges incident to u from S
9: **end while**

A possible execution of this algorithm in a sample graph is shown in Figure 6.13 where the initial vertex included in the cover is f, followed by e, b and g to give a MCVC which is also a minimum connected vertex cover for this graph.

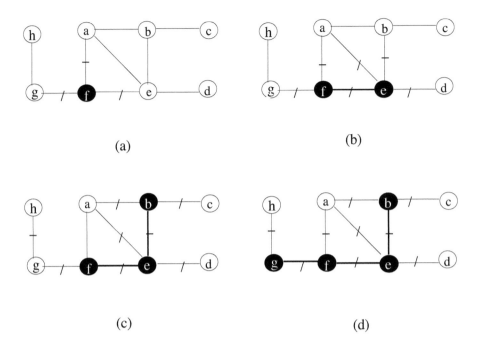

Figure 6.13: *MCVC_Alg* **execution in a sample graph**

6.5.2 A Minimal Weighted Vertex Cover Algorithm

For a vertex weighted graph $G(V, E, w)$, we will attempt to find a minimal weighted vertex cover. We can simply proceed as in the greedy highest degree first algorithm of Section 3.9 and select the vertex that has the minimum weight instead of the highest degree among the vertices. The pseudocode for this algorithm called *MWVC_Alg* is shown in Alg. 6.5.

Algorithm 6.5 *MWVC_Alg*

1: **Input**: A vertex weighted graph $G(V, E, w)$
2: $S \leftarrow E$
3: $v\_list \leftarrow$ sorted list of vertices in ascending weights
4: **while** $S \neq \emptyset$ **do**
5:     $u \leftarrow$ first element of v_list
6:     $MWVC \leftarrow MWVC \cup \{u\}$
7:     **remove** all edges incident to u from S
8:     $v\_list \leftarrow v\_list \setminus \{u \cup \{$all vertices that do not have uncovered edges in $S\}\}$
9: **end while**

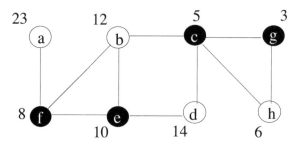

Figure 6.14: *MWVC_Alg* **execution in a sample graph**

Figure 6.14 displays a possible execution of this algorithm in a sample graph where the vertices $\{g, c, f, e\}$ are included in the cover consequently. Vertex h is not included in the cover as it has no uncovered edges left when c is included in the vertex cover.

The time complexity for this algorithm is $O(n)$ as in the case of a linear network. Constructing a minimal weighted and connected vertex cover can be performed similarly and is left as an exercise (see Ex. 7). Our next step is the forming of a distributed minimal vertex cover algorithm as described next.

6.6 A Distributed Algorithm for MWVC Construction

In the case of a computer network, it may be required to have a MVC algorithm at the end of which every node of the network becomes aware whether it is in the cover or not. Such a requirement necessitates the implementation of a distributed algorithm structure of which was described in Section 3.11. The algorithm we will design assumes that there are attributes attached with the nodes in terms of weights which represent a parameter, for example ages in a social network. Our aim is to have each edge covered by the vertices as before, meaning each person in the social network is connected to an aged node in this network but not to each other.

We will assume there is an overlay spanning tree that covers all of the nodes of the network and the algorithm is executed in synchronous rounds initiated by a special node called the *root* as in an SSI algorithm. In each round, if a node u finds it has the highest (or lowest) weight among its neighbors, it decides to enter the MVC and tells its decision to all of its neighbors which can then remove u from their active neighbor list and also remove the edge associated with this neighbor from their uncovered edges list. The algorithm continues until all nodes have all their incident edges covered. A single round of this algorithm called *Dist_MVC* is shown in Alg. 6.6. The synchronization by the *upcast* messages to the *root* is not shown for simplicity.

Algorithm 6.6 *DistMWVC_Alg*

1: **set of int** *uncovd_edges* ← Γ(*i*); *nb_weights* ← neighbor weights as tuples < *node_id*, *weight* >, *received* ← {∅}
2: **message types** *round*, *decide*, *undecide*
3: int *i*, *j*   ▷ *i* is this node, *j* is the sender of a message received
4: **boolean** *covered*
5: **for** *round* = 1 to *n* **do**
6:     **while** *received* ≠ *curr_neighs* **do**   ▷ round *k* for all nodes
7:         **receive** *msg*(*j*)
8:             **case** *msg*(*j*).*type* **of**
9:                 *round*(*k*):   **if** ¬*covered* ∧ (*uncovd_edges* ≠ ∅) **then**
10:                                    **if** *i* < *min*(*w*|{< *j*, *w* >} ∈ *nb_weights*) **then**
11:   **send** *decide*(*k*) to *curr_neighs*
12:   *covered* ← *true*
13:                                    **else send** *undecide*(*k*) to *curr_neighs*
14:                 *decide*(*k*):   *nb_weights* ← *nb_weights* \ {< *j*, *w* >}
15:                                 *received* ← *received* ∪ {*j*}
16:                                 *uncovd_edges* ← *uncovd_edges* \ {*j*}
17:                                 *curr_neighs* ← *curr_neighs* \ {*j*}
18:                 *undecide*(*k*):   *received* ← *received* ∪ {*j*}
19:     **end while**
20: **end for**

The operation of Alg. 6.6 is shown in Figure 6.15 with 8 nodes. We assume that all nodes initially know the weights of their neighbors. Nodes 1 and 2 have the lowest weights in their neighborhoods and they enter the MWVC in the first round of the algorithm. All of their incident edges are removed from the network by their neighbors. The nodes included in the cover are 4 and 3 in round 2, and node 5 in round 3 where the algorithm stops as there are no uncovered edges left. The final MWVC is shown in (d) with five nodes.

Analysis

Theorem 6.2
Alg. 6.6 provides a minimal weighted vertex cover of a graph G in $O(n)$ rounds using $O(nm)$ messages.

Proof 9 *Alg. 6.6 continues until there are no uncovered edges, therefore, the output is a MVC. The linear network where nodes are ordered in increasing or decreasing weights gives us again the worst configuration in which case we need to have $O(n)$ rounds as the algorithm will in fact run sequentially, starting with the lowest weight node. Each edge of the network will be traversed by at most one decide message in*

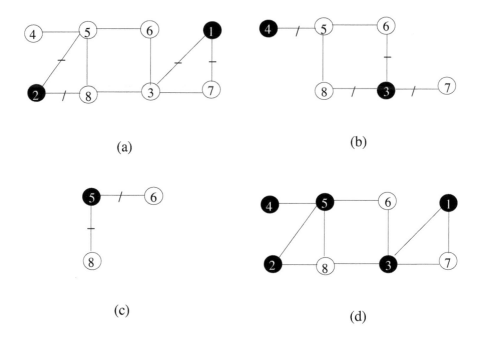

(a)

(b)

(c)

(d)

Figure 6.15: The three rounds of the distributed vertex cover algorithm

---

one direction or at most two undecide messages in both directions resulting in $O(m)$ messages per round. The total number of messages used will then be $O(nm)$.

Alg. 6.6 has the same approximation ratio as in the serial case of $O(m \log n)$ as it basically has the same principle of operation. It is slow as its execution time depends on the number of nodes and for a large network this would not be favorable.

6.7 Chapter Notes

We have reviewed algorithms that construct special subgraphs. The problems encountered are NP-hard in many cases except for the matching problem which can be solved in polynomial time. For this reason, heuristics are frequently used to construct these special subgraphs.

These subgraphs have properties that may be used for complex network applications. For example, it may be required to have a minimum set of administrators in a social network where every person in the network communicates with at least one administrator. The minimal dominating set of this network would provide this function and if the administrators are required to be connected to each other, a minimal connected dominating set will be adequate. Minimal dominating sets are used for clustering around clusterheads where every clusterhead is a member of the MDS and a node that is dominated by the clusterhead is a member of that cluster iedntified by it. Minimal connected dominating sets are also widely used to construct a communication backbone for routing purposes in wireless sensor networks and mobile ad hoc networks.

Maximal independent sets can be used as building blocks for other algorithms such as dominating sets. Matching algorithms are used for the allocation of communication channels in computer networks. As a social network application, bipartite weighted matching algorithms where matching is performed between two groups of people can be employed for forming men and women friendship networks. For the vertex cover application in a social network, we may be interested in forming a minimal group of people with special functions such as social workers where each person is attached to at least one person of this special group but there are not any connections between persons in the large group.

Exercises

1. Show an MIS and an MaxIS of the the sample graph in Figure 6.16.

2. Find an MDS, MCDS and an MinDS of the the sample graph in Figure 6.17.

3. Modify *MDS_Alg* of Section 6.3 so that the output is a connected MDS. Show the operation of this algorithm in the graph of Figure 6.14.

4. Provide a modification to *MDS_Alg* so that each node that is not in the dominating set is at most 2 hops away from at least one node in the dominating set. Work out the time complexity of this algorithm.

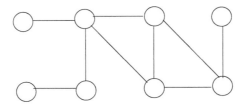

Figure 6.16: Example graph for Ex. 1

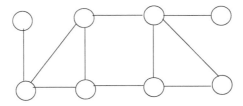

Figure 6.17: Example graph for Ex. 2

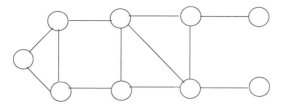

Figure 6.18: Example graph for Ex. 5

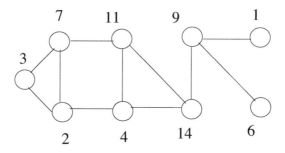

Figure 6.19: Example graph for Ex. 6

5. Figure out an MM, MaxM and MaxWM of the sample graph in Figure 6.18.

6. Show a MVC, MinVC and MinWVC of the sample graph in Figure 6.19 where weights of vertices are shown next to them.

7. Modify the pseudocode for *MWVC_Alg* so that the output is a connected minimal weighted vertex cover. Show the execution of this algorithm in the graph of Figure 6.14.

8. Provide a modification to *DistMWVC_Alg* of Section 6.8 such that the resulting vertex cover is connected. Show the execution of this algorithm in Figure 6.15, prove it works correctly and work out its time and message complexities.

References

[1] D. Avis. A survey of heuristics for the weighted matching problem. *Networks*. 13(4):475-493, 1983.

[2] Erciyes, K. *Distributed Graph Algorithms for Computer Networks*. Computer Communications and Networks, Springer, 2013, ISBN 978-1-4471-5172-2, chapters 9-13.

[3] J. Edmonds. Path, trees, and flowers. *Canadian J. Math.* 17:449-467, 1965.

[4] M.R. Garey and D.S. Johnson. *Computers and Intractability: A Guide to the Theory of NP-Completeness.* Freeman, 1978.

[5] S. Guha and S. Khuller. Approximation algorithms for connected dominating sets. *Algorithmica*, 20(4):374-387, 1998.

[6] E. Halperin. Improved approximation algorithms for the vertex cover problem in graphs and hypergraphs. *SIAM Journal on Computing* 31(5):1608-1625, 2002.

[7] R. Preis. Linear time 1/2-approximation algorithm for maximum weighted matching in general graphs. In *Proceedings of 16th STACS*, C. Meinel and S. Tison (Eds.), LNCS 1563, Springer, 259-269, 1999.

[8] C. Savage, Depth-First search and the vertex cover problem, *Information Processing Letters*, 14(5):233-235, 1982.

[9] R. Wattenhofer. *Principles of Distributed Computing Course*, Chapter 12, Class notes, ETH Zurich, 2004.

Chapter 7

Data Clustering

7.1 Introduction

A cluster is a collection of objects that are similar to each other using some attribute and therefore can be treated as a group. Some examples of objects under consideration are biological data points obtained experimentally, people in social networks and mobile ad hoc computers. Output of a clustering method should provide clusters where intra-cluster similarity is high and inter-cluster similarity is low. The quality of a clustering method is dependent on the similarity measure and the algorithm employed.

Formally, given a set $P = \{p_1, ..., p_n\}$ of n data points representing n objects, the goal of clustering is to divide P into k groups $C_1, ..., C_k$ such that data belonging to a group are more similar to each other than data in other groups. Each C_i is called a *cluster* and the number of clusters may be input by the user, or the algorithm may determine it as its output. Clusters may overlap, meaning a data point may belong to more than one cluster or they may be disjoint with no common elements.

Clustering has numerous applications including bioinformatics data analysis, parallel processing, image processing, psychology, business and document classification in the World Wide Web. There are various clustering algorithms with different goals and the general requirements from any clustering algorithm are that it should be scalable, that is, its performance should not degrade as data size gets larger; it should be able to cluster data groups with arbitrary shapes, and it should also consider noise data points and outliers. Data points may have different attributes and the algorithm should be able to deal with these attributes. Also, it is preferred to have a minimum number of user-specified inputs to the algorithm and the clustering output should not change when data set is duplicated and clustering algorithm is executed again.

The clustering task is considered to have the following steps [15]: feature selection or extraction; clustering algorithm design or selection; validation of clusters; and interpretation of results. In the first step, desirable features of clustering are selected and an algorithm suitable for these features is determined in the second step. The validation of the clusters is tested using internal, external and relative cluster validity. Finally, the user is presented with the cluster structure in a usable format in the last step.

There is not a single standard clustering algorithms for all applications and even the same clustering algorithm may have different performances with different data sets. Clustering of data points can be performed by linear clustering and non-linear clustering methods. Hierarchical clustering algorithms, k-means clustering and fuzzy clustering algorithms are examples of linear algorithms; and density based clustering and minimum spanning tree based clustering algorithms are examples of non-linear clustering algorithms. In this chapter, we will investigate representative examples of these algorithmic methods of data clustering by first describing the types of data clustering.

7.2   Types of Data Clustering

Clustering can be performed using various methods such as hierarchical or partitional; overlapping, disjoint or fuzzy; and complete or partial [14].

Hierarchical or Partitional Clustering

An important distinction between the clustering methods is whether a cluster can be included in another cluster. In *hierarchical clustering*, the clusters are nested and can be displayed as a tree. Each cluster then contains all of the nodes including its root node in a subtree. On the other hand, the objects are divided into non-overlapping, therefore unnested clusters in *partitional clustering*.

Disjoint, Overlapping or Fuzzy Clustering

Another distinction is made about the assigned membership of the nodes in the clusters at the end of the clustering algorithm. In *disjoint clusters*, each node is a member of exactly one cluster whereas a node may be a member of more than one cluster in *overlapping clusters*. Fuzzy clustering methods assign a membership weight between 0 and 1 to each node for every cluster obtained such that 1 means absolute membership, 0 means a non-member and any weight between 0 and 1 shows the probability of the node to be in that cluster.

Data or Graph-theoretic Clustering

A distinction can be made about clustering data points or objects that are represented as a graph. In *data clustering*, data points are placed in a 2D plane and the aim is to cluster these points based on a metric such as the distance between these points. If the objects can be represented as a graph $G(V,E)$ where they are the vertices of G and the relationship between them, which is not necessarily the distance between them, is represented by the edges of G; we can use graph-theoretic concepts and algorithms to perform clustering. In this chapter, we will mainly describe data clustering algorithms and defer discussion of graph-based clustering algorithms to Chapter 8.

7.3   Agglomerative Hierarchical Clustering

Hierarchical clustering algorithms are basically of two types, *agglomerative* or *divisive* algorithms. Agglomerative hierarchical algorithms start by each node representing a single cluster (singleton) and group the nodes based on their distances to each other. At each iteration, the distance between the clusters are recalculated and merging operation is performed. The divisive algorithms on the other hand, start with one cluster which includes all of the nodes and iteratively divide the clusters into smaller ones.

The output of a hierarchical clustering algorithm is a tree-like diagram called a *dendogram* which shows the nested clusters as shown in Figure 7.1 where five points $p_1, .., p_5$ are clustered using agglomerative hierarchical clustering which considers shortest distance between the two closest points in each pair of clusters. The leaves of the dendogram represent individual points, the number of iterations in sequence are shown in italic numbers and horizontal lines divide the dendogram into the required number of clusters. The line L_1 divides the dendogram into two clusters $\{p_1, p_4\}$ and $\{p_3, p_5, p_2\}$ and L_2 divides it into three clusters $\{p_1, p_4\}$, $\{p_3, p_5\}$ and $\{p_2\}$ as shown.

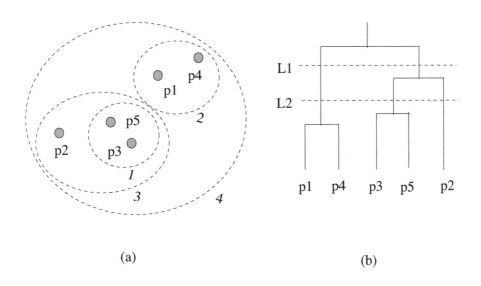

(a)   (b)

Figure 7.1: a) The nested clusters. b) The corresponding dendogram

The agglomerative hierarchical clustering algorithms consist of the steps shown in Alg. 7.1.

Algorithm 7.1 *Agglomerative Hierachical Clustering*

1: **Input** : $P = \{p_1, p_2, ..., p_n\}$ ▷ n data points to be clustered
2: **Output** : A dendogram showing the cluster structure
3: **compute** the proximity matrix for data points
4: **repeat**
5:     **merge** the two closest clusters into one cluster
6:     **update** the proximity matrix
7: **until** there is only one cluster

The agglomerative hierarchical clustering algorithms consider the shortest distance between all pairs of clusters during the merge operation. The definition of the distance between the clusters varies between various algorithms. The cluster proximity can be defined using one of the following metrics:

1. *Single linkage*: The shortest distance between the two closest points in every pair of clusters are computed.

2. *Complete linkage*: The shortest distance between the two farthest points in every pair of clusters are computed.

3. *Average linkage*: The average distance between the data points for every pair of clusters are found.

The single-linkage, complete-linkage and the average linkage methods are shown in Figure 7.2.

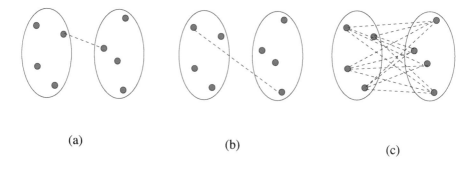

(a)   (b)   (c)

Figure 7.2: a) Single linkage. b) Complete linkage. c) Average linkage

The single link agglomerative hierarchical clustering algorithm is one of the most commonly used hierarchical clustering algorithms. The implementation code for this algorithm is shown in detail in Alg. 7.2 (*AHC_Alg*) where we have a distance matrix $D[n,n]$ which shows the current distances between the clusters formed and is initialized to the distances between the data points as each data point is a cluster initially. Then, at each iteration, we find the smallest value in this matrix corresponding to the

---

Algorithm 7.2 *AHC_Alg*

1: **Input :set of points** $P = \{p_1, p_2, ..., p_n\}$ ▷ n data points to be clustered
2:   $D[n,n]$ ← distances between clusters
3: **Output** : set of embedded clusters $C = \{C_1, ..., C_n\}$
4: **int** m ← n ▷ size of D which decreases at each iteration
5: **for** i ← 1 to n **do**
6:     C_i ← $\{p_i\}$ ▷ each point is a cluster initially
7: **end for**
8: **while** $m > 1$ **do**
9:     min_val ← $D[1,1]$ ▷ find minimum distance value
10:     **for** i ← 2 to m **do**
11:        **for** j ← $i+1$ to m **do**
12:           **if** $i \neq j \wedge D[i,j] < min\_val$ **then**
13:             min_val ← $D[i,j]$
14:             a ← $min(i,j)$; b ← $max(i,j)$ ▷ find lower and higher indices
15:           **end if**
16:        **end for**
17:     **end for**
18:     **for** i ← 1 to m **do**   ▷ update distances
19:        **if** $i \neq a$ **then**
20:           $D[i,j]$ ← $min(D[i,a], D[i,b])$
21:           $D[j,i]$ ← $D[i,j]$
22:        **end if**
23:     **end for**
24:     **for** j ← b to m **do**   ▷ shift columns
25:        **for** i ← 1 to m **do**
26:           $D[i,j]$ ← $D[i,j+1]$
27:        **end for**
28:     **end for**
29:     **for** i ← b to m **do**   ▷ shift rows
30:        **for** j ← 1 to m **do**
31:           $D[i,j]$ ← $D[i+1,j]$
32:        **end for**
33:     **end for**
34:     C ← $C_a \cup C_b$ ▷ merge clusters
35:     m ← $m - 1$
36: **end while**

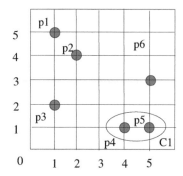

	p_1	p_2	p_3	p_4	p_5	p_6
p_1	0	$\sqrt{2}$	3	5	$\sqrt{32}$	$\sqrt{20}$
p_2		0	$\sqrt{5}$	$\sqrt{13}$	$\sqrt{18}$	$\sqrt{10}$
p_3			0	$\sqrt{10}$	$\sqrt{17}$	$\sqrt{17}$
p_4				0	1	$\sqrt{5}$
p_5					0	2
p_6						0

Figure 7.3: Hierarchical algorithm first iteration

shortest distance between the closest points in all pairs of clusters. The entry for this distance has a as the lower index and b as the higher index. We update distances to the new cluster formed by merging these two clusters and then shift rows and columns of D to prepare it for the next iteration. This process is repeated until we have only one cluster that contains all of the clusters.

Figures 7.3 through 7.6 display the operation of the *AHC_Alg* in a data set consisting of six points $p_1, .., p_6$. The shortest distance in the first iteration is between the points p_4 and p_5 which are combined into the cluster C_1 and the distances from all clusters to this cluster are updated in the matrix D. The iterations continue until we have one cluster C_5 which includes all of the clusters $C_1, .., C_4$.

Figure 7.7 displays the dendogram obtained by the hierarchical clustering algorithm. This dendogram can be intersected by a horizontal line to yield the required number of clusters. For example, the line A divides the dendogram into three clusters $C_1 = \{p_1, p_2\}$, $C_2 = \{p_3\}$ and $C_2 = \{p_4, p_5, p_6\}$ whereas line B provides four clusters $C_1 = \{p_1, p_2\}$, $C_2 = \{p_3\}$, $C_3 = \{p_4, p_5\}$ and $C_6 = \{p_6\}$.

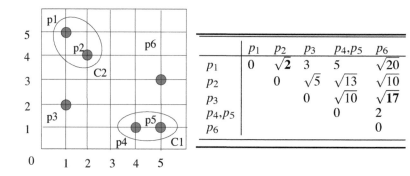

	p_1	p_2	p_3	p_4,p_5	p_6
p_1	0	$\sqrt{2}$	3	5	$\sqrt{20}$
p_2		0	$\sqrt{5}$	$\sqrt{13}$	$\sqrt{10}$
p_3			0	$\sqrt{10}$	$\sqrt{17}$
p_4,p_5				0	2
p_6					0

Figure 7.4: Hierarchical algorithm second iteration

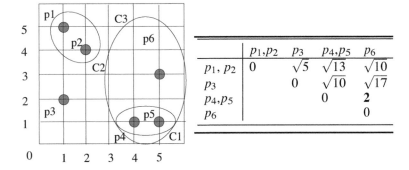

	p_1, p_2	p_3	p_4, p_5	p_6
p_1, p_2	0	$\sqrt{5}$	$\sqrt{13}$	$\sqrt{10}$
p_3		0	$\sqrt{10}$	$\sqrt{17}$
p_4, p_5			0	2
p_6				0

Figure 7.5: Hierarchical algorithm third iteration

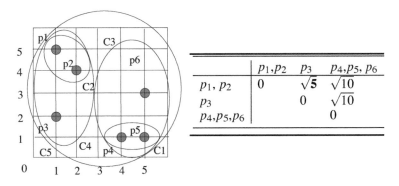

	p_1, p_2	p_3	p_4, p_5, p_6
p_1, p_2	0	$\sqrt{5}$	$\sqrt{10}$
p_3		0	$\sqrt{10}$
p_4, p_5, p_6			0

Figure 7.6: Hierarchical algorithm fourth iteration

Each iteration of the agglomerative clustering algorithm requires $O(n^2)$ time to find the shortest distance in matrix D, and this process is repeated for at most n times resulting in a time complexity of $O(n^3)$. Using suitable data structures, time complexity can be reduced to $O(n^2 \log n)$. The hierarchical algorithms are easy to implement and may provide the best results in some cases. Another advantage of these algorithms is that the number of clusters is not needed beforehand. However, these algorithms are not sensitive to noise data points and have difficulty in discovering arbitrarily shaped clusters. They also do not directly optimize an objective function.

7.4 k-means Algorithm

The k-means algorithm is a simple and effective clustering algorithm which assumes the number of clusters (k) to be formed is known *a priori*. Initially, k centroids are chosen randomly to form k clusters. The initial placement of these centers may have

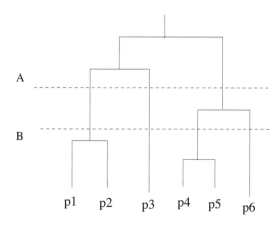

Figure 7.7: Dendogram showing clusters of Figure 7.6

a profound effect on the structure of the clusters to be formed. Each data point is then assigned to its nearest centroid and data points assigned to the same centroid form a cluster. In the next step, the centroids for each cluster are recomputed and the points are again assigned to their nearest centroids, forming new clusters. This process continues until the coordinates of the centroids (or the membership of data points) do not change significantly as shown in the basic structure of Alg. 7.3.

Algorithm 7.3 k-means Algorithm Basic Structure

1: **Input** : $P = \{p_1, p_2, ..., p_n\}$ ▷ n data points to be clustered
2: **Output** : k clusters
3: **select** k points as the initial centroids
4: **repeat**
5:     **assign** each data point to its closest centroid and form clusters
6:     **compute** the centroid for each cluster
7: **until** centroids do not change significantly

The Initial Positioning of Centroids

Positioning of the centroids at random may result in different clusters for each run of the algorithm. One method to overcome this problem is to use hierarchical clustering initially to obtain the clusters and use the centroids of these clusters as the initial centroids of the k-means algorithm. If the sample size is not large and k is relatively small with respect to the sample size, this method provides favorable results.

The Objective Function

Computing the centroid of each cluster in this algorithm is crucial for the quality of the clusters produced. As an objective function, sum of the squared error (SSE) may be used. In order to obtain SSE, the Euclidean distance of each data point to its centroid called the *error* of the point is calculated and the total sum of the errors is computed in the second step. We then compare the results obtained by different runs of the k-means algorithm and use the output that gives the smallest SSE of the centroids for this run as it has a better approximation of the centroids [14]. The formal definition of SSE is as follows:

$$SSE = \sum_{i=1}^{k} \sum_{p \in C_i} d(p, c_i)^2 \tag{7.1}$$

where C_i are the clusters, c_i are the centroids of the clusters, p are the points in the clusters, and $d(p, c_i)$ is the distance between the data points and the centroids. The centroid of a cluster that minimizes the SSE can then be specified as the mean of the data points within the cluster as:

$$c_i = \frac{1}{n_i} \sum_{p \in C_i} p \tag{7.2}$$

In this case, we can simply find the arithmetic average of the x coordinates of all data points within a cluster to find the x coordinate of the centroid and similarly the arithmetic average of all y data point coordinates yields the y coordinate of the centroid for that cluster. A possible detailed and implementation oriented code for this algorithm is shown in Alg. 7.4 where a data set of n points $p_1, ..., p_n$ is to be clustered; $C_1, .., C_k$ are the dynamic clusters which store the points with dynamic cluster centroids $c_1, ..., c_k$ respectively. The stopping condition for this algorithm is when the coordinates of the centroids do not change significantly in an iteration. In order to detect this condition, we sum the x and y coordinates of the centroids $c_j(x_j, y_j)$. When the sums of these coordinates computed in the current iteration vary less than a threshold value τ from the sums of the last iteration, the algorithm stops. We start by assigning k centroids at random initially and start each iteration by computing the sum of the coordinates for the newly found centers first. We then store the distances between each data point and each cluster centroid in the distance matrix $D[n, k]$. In the next step, each data point is included in the cluster of its closest centroid and the that number of elements of that cluster is incremented.

We are now ready to find the coordinates of the new centroids based on the coordinates of the newly placed points in the clusters. In this case, we simply find the arithmetic means of the x coordinates of all data points in a cluster to find the x coordinate of the new centroid for that cluster and obtain y coordinate of the centroid similarly. For these new centroids, the summation of their x and y coordinates is done to check the stopping condition.

Algorithm 7.4 $k - means\_Alg$

1: **Input :set of points** $P \leftarrow \{p_1(x_1,y_1),...,p_n(x_n,y_n)\}$ ▷ n data points
2: **Output : set of clusters** $C \leftarrow \{C_1,...,C_k\}$ ▷ k clusters
3: **set of points** $c_1,..,c_k$ ▷ k centroids
4: **real** $D[n,k]$ ▷ distances between k cluster centers and n points
5: **int** $n\_pts[k]$ ▷ array for number of points in each cluster
6:
7: **assign** k random centers $c_1,...,c_k$ initially
8: **for** $j = 1$ to k **do**   ▷ find sums of centroid coordinates
9:      $XT_{new} \leftarrow XT_{new} + c_j(x_j)$
10:      $YT_{new} \leftarrow YT_{new} + c_j(y_j)$
11: **end for**
12: **repeat**
13:      **for** $j = 1$ to k **do**
14:          $XT \leftarrow XT_{new}$; $YT \leftarrow YT_{new}$ ▷ update centroid coordinates
15:          $C_j \leftarrow \emptyset$ ▷ empty clusters
16:          **for** $i = 1$ to n **do**
17:              $D[p_i,c_1] \leftarrow d(p_i,c_j)$ ▷ store distances to new centroids in D
18:          **end for**
19:      **end for**
20:      **for** $i = 1$ to n **do**   ▷ find nearest centroid
21:          $dist \leftarrow D[p_i,c_1]$
22:          **for** $j = 1$ to k **do**
23:              **if** $D[p_i,c_j] \leq dist$ **then**
24:                  $dist \leftarrow d(p_i,c_j)$; $ind \leftarrow j$
25:              **end if**
26:          **end for**
27:          $C_{ind} \leftarrow C_{ind} \cup \{p_i\}$ ▷ include p_i in nearest cluster
28:          $n\_pts[j] \leftarrow n\_pts[ind] + 1$ ▷ increment point count in the cluster
29:      **end for**
30:      $XT_{new} \leftarrow 0$; $XT_{new} \leftarrow 0$
31:      **for** $j = 1$ to k **do**
32:          $X_j = \sum_{p_i(x_i,y_i) \in C_j} x_i$ ▷ sum x and y coordinates of points
33:          $Y_j = \sum_{p_i(x_i,y_i) \in C_j} y_i$
34:          $c_j(x_j) \leftarrow X_j / n\_pts[j]$ ▷ find new centroid x and y coordinates
35:          $c_j(y_j) \leftarrow Y_j / n\_pts[j]$
36:          $XT_{new} \leftarrow XT_{new} + c_j(x_j)$; $YT_{new} \leftarrow YT_{new} + c_j(y_j)$
37:      **end for**
38: **until** $|XT_{new} - XT| + |YT_{new} - YT| \geq \Theta$ ▷ check differences

---

Figure 7.8 displays the execution of the k-means algorithm in a data set of eleven points $p_1,...,p_{12}$. The initial placement of random centers are shown in (a) and we see that only after three iterations, the newly computed centers are stable and the algorithm stops.

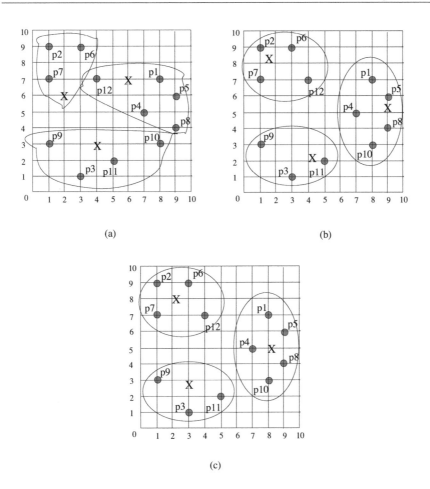

Figure 7.8: *k*-**means algorithm execution example. a) The initial centers and clusters. b) First iteration. c) Stable centers and clusters**

Time Complexity

The *k*-means algorithm has a time complexity of $O(tknd)$ where n is number of objects under consideration, k is number of clusters, d is number of attributes for each data point and t is number of iterations. Normally, $k, t, d << n$, we can therefore assume that this algorithm is efficient and converges fast. It provides reasonable clusters when data points are well separated from each other. However, there are some disadvantages; firstly, it requires the number of clusters to be specified beforehand. Secondly, the placement of the initial centroids may effect the cluster structure and we need to have additional procedures for proper placement of the centroids as described before. This algorithm fails to detect overlapping clusters and non-linear data sets. It also fails to handle noisy data and outliers.

7.5 Nearest Neighbor Algorithm

The nearest neighbor clustering algorithm proposed by Lu and Fu [12] forms clusters of data points based on their distances to their closest neighbors. This algorithm iteratively checks each data point and assigns it to the cluster of its nearest neighbor if the distance between these two data points is below a given threshold. A possible implementation of this procedure is shown in Alg. 7.5 where the closest neighbor p_m of the point p_i under consideration is found and if $d(p_i, p_m)$ is below a threshold τ, p_i is included in the cluster of p_m. Otherwise, p_i is included in a new cluster.

Algorithm 7.5 *NNC_Alg*

 1: **Input** : set of points $P = \{p_1, ..., p_n\}$ ▷ n data points
 2: **Output** : set of clusters $C = \{C_1, ..., C_k\}$ ▷ clusters
 3: **real** $D[n, n]$ ▷ distances between n data points
 4: $C_1 \leftarrow \{p_1\}$ ▷ p_1 is initialized with first cluster
 5: **for all** $p_i \in P$ **do**
 6:     **find** $p_m \in C_m$ such that $D[p_i, p_m]$ is the smallest
 7:     **if** $D[p_m, p_i] < \tau$ **then**   ▷ check if distance is smaller than threshold
 8:         $C_m \leftarrow C_m \cup \{p_i\}$
 9:     **else**
10:         $k \leftarrow k + 1$
11:         $C_k \leftarrow p_i$ ▷ start a new cluster
12:     **end if**
13: **end for**

The execution of this algorithm for twelve data points $p_1, ..., p_{12}$ is shown in Figure 7.9 and six clusters are obtained as a result, using a threshold distance value of $\sqrt{5}$. The nearest neighbor clustering algorithm has two search the distance matrix D in $O(n)$ time at each iteration for a total of n points, therefore, it has $O(n^2)$ time complexity.

7.6 Fuzzy Clustering

Clustering algorithms we have seen up to now provide non-overlapping clusters, a process sometimes referred as *exclusive clustering* where each data point belongs to exactly one cluster. This type of clustering is called *hard clustering* in contrast to *fuzzy clustering* where each data point has a membership value which is typically a real number between and including 0 and 1 for each cluster [17] . The membership matrix $U[n, k]$ has an element $u_{ij} \in [0, 1]$ which represents the membership value for object x_i in cluster C_j. For example, given four clusters $C_1, ..., C_4$ and 20 data points $p_1, ..., p_{20}$, the the 12th row of U may have 0.2, 0.8, 0.3, 0.1 values which shows that the membership value of point p_{12} is much higher in C_2 than the other clusters.

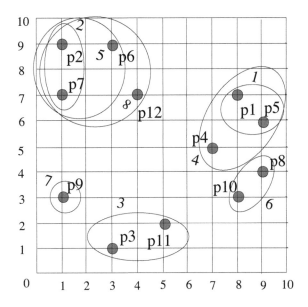

Figure 7.9: Nearest neighbor algorithm execution. The clusters formed in sequence are shown by italic numbers

A higher membership value indicates a higher confidence in the assignment of the object to the cluster. Figure 7.10 shows an example of randomly scattered 10 data points in 2D plain.

One possible fuzzy partitioning for this data set assuming two clusters C_1 and C_2 are required may form the matrix U as follows:

$$U = \begin{bmatrix} 1.0 & 1.0 & 0.7 & 0.9 & 0.6 & 0.5 & 0.3 & 0 & 0 & 0.2 \\ 0 & 0 & 0.3 & 0.1 & 0.4 & 0.5 & 0.7 & 1 & 1 & 0.8 \end{bmatrix}$$

We have included data points that have a membership value equal or greater than 0.5 and as such, there are two fuzzy clusters C_1 and C_2 formed. We can see that points p_1 and p_2 are definitely in cluster C_1, and the points p_8 and p_9 are in cluster C_2 as their membership values for these clusters are 1. The rest of the points have cluster membership values between 0 and 1 and the sum of their membership values is always equal to unity. For example, point p_5 is slightly closer to cluster C_1 elements than C_2 and it has membership values 0.6 and 0.4 for these two clusters, respectively. Point p_6 is equidistant to both clusters and has 0.5 membership values for both clusters.

A general fuzzy clustering algorithm starts by an initial fuzzy partition of n objects into k clusters specified by the membership matrix U. It then evaluates a fuzzy criterion function and reassigns objects to clusters to improve this function. This process is repeated until the entries in U are stable [14]. A widely used fuzzy clustering

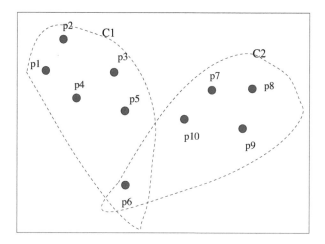

Figure 7.10: Fuzzy clustering example.

algorithm is called *fuzzy c-means algorithm* (FCM) [1] which assigns membership to each data point corresponding to each cluster center on the basis of distance between the cluster center and the data point. The closer the data point is to the cluster center, the stronger its membership in that cluster. Clearly, summation of membership of each data point should be equal to one. Alg. 7.6 shows one way of implementing FCM.

The parameter m in the algorithm is the fuzziness index, d_{ij} is the Euclidean distance between ith data and jth cluster center, Θ is the termination criterion between [0,1] and x is the iteration step. It computes fuzzy membership u_{ij} first and then finds the fuzzy centers using the membership values. The algorithm continues until fuzzy membership values do not change anymore. In this respect, it has a similar structure to k-means algorithm but the membership values of the points are not hard. The FCM algorithm has favorable performance for overlapping data points but we need to specify the number of clusters beforehand as in k-means algorithm. Lower values of Θ result in better precision at the expense of increased number of iterations. In order to speed up the calculation of c-means, the data set is represented as a histogram in [2] and the exact conditions are replaced with approximate ones in [3] to allow table looking up.

7.7 Density-based Clustering

Density-based clustering algorithms use local connectivity and density functions to perform clustering. Density of points inside a cluster should be significantly higher than the density of points outside the cluster after running these algorithms. The advantages of density-based algorithms are that the clusters discovered may have

---

Algorithm 7.6 *FCM_Alg*

1: $P = \{p_1, p_2, ..., p_n\}$ ▷ *n* data points to be clustered

2: $C \leftarrow \{c_1, c_2, ..., c_q\}$ ▷ *q* cluster centers

3: **repeat**

4:      **select** *q* cluster centers at random

5:      **calculate** fuzzy membership u_{ij} using:

$$u_{ij} \leftarrow 1 / \sum_{k=1}^{q} (d_{ij}/d_{ik}^2/m - 1)$$

6:      **compute** the fuzzy centers c_j using:

$$c_j \leftarrow (\sum_{i=1}^{n} u_{ij}^m p_i) / (\sum_{i=1}^{n} u_{ij}^m), \forall j = 1, 2, ..., q$$

7: **until** $|U^{x+1} - U^x| < \Theta$

---

arbitrary shapes as shown in Figure 7.11, and noise is also handled. Examples of density based algorithms are DBSCAN, GDBSCAN, OPTICS and DENCLUE.

Two parameters that are usually considered in these algorithms are the maximum radius of the neighborhood (*Eps*) and the minimum number of points in the neighborhood bounded by *Eps* (*MinPts*). Based on these parameters, types of data point considered are as follows:

- *Core points*: Objects that are within the *Eps* radius and have at least *MinPts* points.

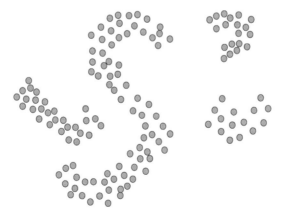

Figure 7.11: Arbitrary clusters obtained by a density-based algorithm

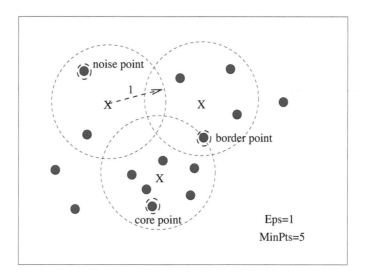

Figure 7.12: Core, border and noise points in density-based clustering

- *Border points*: Points that are on the border of a cluster

- *Noise points*: Points that are neither core nor border points.

Figure 7.12 displays these three types of data points as a core point and a border point of a cluster and noise point. The *MinPts* is equal to 5 and the *Eps* is unity in this example.

Before describing the representative density-based algorithm called the density-based spatial clustering of applications with noise (DBSCAN), we will have the following definitions.

Definition 8 directly density-reachable *A point q is a neighbor of a point p with respect to Eps if $d(p,q) \leq Eps$. Such neighbors of p are shown as $N_{Eps}(p)$. A point q is directly density-reachable from a point p with respect to Eps and MinPts if $q \in N_{Eps}(p)$ and $|N_{Eps}(p)| \geq MinPts$.*

Definition 9 density-reachable *A point q is density-reachable from a point p with respect to Eps and MinPts if there is a sequence of points $p, p_1, ..., p_k, q$ such that any point p_i in this sequence is directly density-reachable from the previous point p_{i-1}.*

Definition 10 density-connected *A point q is density-connected to a point p with respect to Eps and MinPts if there exists a point r such that p and q are both density-reachable from r with respect to Eps and MinPts.*

We can now look into the operation of the DBSCAN algorithm which classifies each data point as either a core, border or noise point as shown in Alg. 7.7.

Algorithm 7.7 *DBSCAN_Alg*

1: **set of points** $P = \{p_1, ..., p_n\}$ ▷ n data points
2: **states** *core, border, noise*   ▷ states of points
3: **states** *Pstates*[n]   ▷ array holding the states of points
4: **Output** : set of clusters $C = \{C_1, ..., C_k\}$
5: $Q \leftarrow P$ ▷ Q is the working set of points
6: **while** $Q \neq \emptyset$ **do**
7:     **pick** $p_i \in Q$
8:     **find** all *density − reachable* points from p_i w.r.t. *Eps* and *MinPts*.
9:     **if** p_i is a core point **then**
10:         *Pstates*[i] $\leftarrow$ *core*   ▷ a cluster is discovered
11:     **else if** p_i is a border point **then**
12:         *Pstates*[i] $\leftarrow$ *border*
13:     **end if**
14:     $C_i \leftarrow$ all core points
15:     $Q \leftarrow Q \setminus C_i$
16:     $i \leftarrow i + 1$
17: **end while**
18: **for all** $p_i \in Q$ with *Pstates*[i] $\neq$ *core* $\wedge$ *Pstates*[i] $\neq$ *border* **do**
19:     *Pstates*[i] $\leftarrow$ *noise*
20: **end for**

In essence, this algorithm includes all core points that are within *Eps* distance of each other in the same cluster. The time complexity of DBSCAN algorithm is $O(n^2)$ as n points will be considered and for each data point, $O(n)$ time will be needed at most to find points in the *Eps* neighorhood. However, using suitable data structures such as *kd* trees, time complexity can be reduced to $O(n \log n)$ [14].

Choice of the values for *Eps* and *MinPts* clearly has an important effect on the clusters and clustering quality obtained. An experimental approach to determine these parameters would be to first determine the distance from a point p_i to its k nearest neighbors denoted by d_k. The d_k values for all data points can be calculated for some value of k and these distance values can be sorted in increasing order and plotted against the number of points. When we see a sharp increase of the distances in this graph, the k value corresponds to a convenient value of *Eps* and this k value taken for *MinPts* will provide core points that are fewer than *MinPts* points [14].

7.8 Parallel Data Clustering

Data clustering demands high computational power due to the size and high dimension of data. For this reason, many parallel data clustering algorithms have been proposed in literature. The k-means algorithm required calculation of the distances between the data points and the current centers at each iteration, and then assignment of the data points to their closest centers. Either distance calculation or assignment to centers can be performed in parallel as shown in [9] and [8]. In the parallel k-means algorithm and the parallel hierarchical algorithm proposed by Li and Fang, $O(n\log n)$ time on the hypercube and $O(n\log^2 n)$ times were obtained in n-node hypercube and n-node butterfly architectures, respectively [11]. The PBIRCH method described in [7] by Garg et al. provides a method to perform k-means clustering in parallel using the SPMD model. Data is distributed to processors and a special processor selects k centers initially which is broadcast to all processors. Each processor then performs clustering in parallel and computes its local centers which are exchanged at the end of a round to find new centers. Recently, Kraj et al. proposed a parallel k-means algorithm called *ParaKMeans* which performs cluster assignment in parallel [10].

For parallelization of hierarchical clustering, Rasmussen and Willett described parallel implementations of the single link clustering on an SIMD array processor [13] and provided a significant constant speedup factor. Olson provided parallel implementations of hierarchical clustering algorithms in $O(n)$ time using the CRCW PRAM model with n nodes, and $O(n\log n)$ time using $n/\log n$ node butterfly network or tree structure. Xu et al. provided the parallel version of DBSCAN called PDBSCAN using a distributed spatial index structure called dR* tree [20]. PDBSCAN uses a master-slave model of parallel computing and first partitions data into k sets where k is the number of available processing elements. Partitioned data is distributed to processing elements which perform clustering in parallel and at the end of this step, clusters are combined where necessary, to get the total clustering structure of data.

7.9   Chapter Notes

Clustering of data points is a thoroughly studied research topic that resulted in many effective algorithms. The surveys by Jain et. al. [5] and Tan et. al. [14] are good starting points for potential researchers on this topic. There is not a single perfect algorithm that solves all of the problems encountered which can be used for clustering in all applications. The initial feature selection of data step of clustering is crucial in determining the type of clustering algorithm to be employed as a certain clustering method may be appropriate for a class of data where some other method may not provide favorable results on the same data set.

In this chapter, we have described fundamental methods of clustering which are partitioning-based, hierarchical, fuzzy and density-based algorithms by leaving the discussion of graph-based clustering algorithms to the next chapter. We have seen commonly used sample algorithms for these types of clustering such as agglomerative hierarchical, k-means, DBSCAN, and fuzzy c-means algorithms and the general conclusion is that each algorithm has its own merits and demerits. For example, if our application involves discovering irregular data patterns of high volume, DBSCAN or its derivative algorithms can be a good choice whereas for regular shaped data with an initial guess of the number of clusters that can be formed, k-means would perform well.

Due to the intensive computation involved, parallel clustering algorithms can be employed. We have briefly surveyed the available parallel algorithms for data clustering and there is still a need for efficient parallel clustering algorithms especially for clustering in complex networks where data set is huge in many cases.

Exercises

1. Show the execution of agglomerative hierarchical algorithm in the sample data set of the graph of Figure 7.13 by showing the modification to the distance matrix at each iteration. Provide also the output dendogram. Where can a horizontal line be drawn in this dendogram to have four clusters?

2. Show the iterations of the k-means algorithm in in the sample data set of the graph of Figure 7.14 graphically for $k = 4$.

3. Show the implementation of the hierarchical divisive algorithm using maximum linkage by showing the steps of the execution and the array of distances in the graph of Figure 7.15.

4. Compare the hierarchical, k-means, nearest neighborhood clustering algorithm in terms of the time complexities.

5. Show the iterations of the nearest neighbor algorithm in in the sample data set of the graph of Figure 7.16 graphically for a threshold value of $\sqrt{5}$.

6. Discuss the methods used for parallelization of k-means and the agglomerative hierarchical algorithms. Can the fuzzy c-means algorithm be parallelized similar to k-means algorithm?

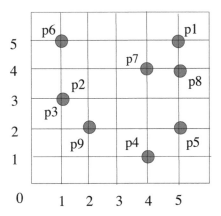

Figure 7.13: An example graph for Ex. 1

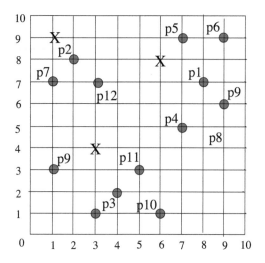

Figure 7.14: An example graph for Ex. 2

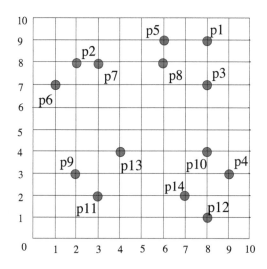

Figure 7.15: An example graph for Ex. 3

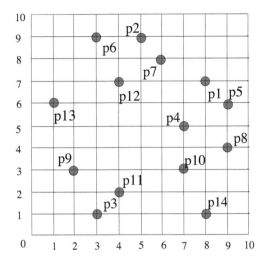

Figure 7.16: An example graph for Ex. 5

References

[1] J. C. Bezdek. *Pattern Recognition with Fuzzy Objective Function Algorithms.* Plenum Press, 1981.

[2] J. C. Bezdek, J. Keller, R. Krisnapuram, and M. Pal. *Fuzzy Models and Algorithms for Pattern Recognition and Image Processing.* Boston, MA: Kluwer, 1999.

[3] R. L. Cannon, J. V. Dave, and J. C. Bezdek. Efficient implementation of the fuzzy c-means clustering algorithm. *IEEE Trans. Pattern Anal. Mach. Intell.,* PAMI-8:248-255, 1986.

[4] M. Ester, H. P. Kriegel, J. Sander and X. Xu. A density-based algorithm for discovering clusters in large spatial databases with noise. In *Proceedings of 2nd International Conference on Knowledge Discovery and Data Mining* (KDD-96), 1996.

[5] A. K. Jain, M. N. Murty and P. J. Flynn. Data clustering: a review. *ACM Computing Surveys,* (31)3: 264-323. 1999.

[6] Cormen, Thomas H., Leiserson, C. E., Rivest, R. L. (1990) *Introduction to Algorithms* (1st ed.). MIT Press and McGraw-Hill. ISBN 0-262-03141-8., Section 26.2, 558-565.

[7] A. Garg, A. Mangla, N. Gupta, and Vasudha Bhatnagar. PBIRCH: a scalable parallel clustering algorithm for incremental data. In *Proceedings of Database Engineering and Applications Symposium,* 315-316, 2006.

[8] D. Judd, P. K. McKinley, and A. K. Jain. Large-scale parallel data clustering. *IEEE Trans. Pattern Anal. Mach. Intell.*, 20(8):871-876, 1998.

[9] S. Kantabutra and A.L. Couch. Parallel K-means clustering algorithm on NOWs. *NECTEC Technical Journal*, 1(1), 1999.

[10] P. Kraj, A. Sharma, N. Garge, R. Podolsky, and Richard A. Mcindoe. ParaK-means: implementation of a parallelized K-means algorithm suitable for general laboratory use. *BMC Bioinformatics*, 9:200, 2008.

[11] X. Li and Z. Fang. Parallel clustering algorithms. *Parallel Computing*, 11:275-290, 1989.

[12] S-Y. Lu and K. S. Fu. A sentence-to-sentence clustering procedure for pattern analysis. *IEEE Trans. Syst. Man Cybern.*, 8:381-389, 1978.

[13] E.M. Rasmussen and P. Willett. Efficiency of hierarchical agglomerative clustering using the ICL distributed array processor. *Journal of Documentation*, 45(1):1-24, 1989.

[14] P-N Tan, M. Steinbach, and V. Kumar, Introduction to data mining, Chapter 8. *Cluster Analysis: Basic Concepts and Algorithms.* Addison-Wesley (1st ed.), ISBN-10: 0321321367, ISBN-13: 978-0321321367, 2005.

[15] R. Xu and D. Wunsch II. Survey of clustering algorithms. *IEEE Transactions on Neural Networks*, 16(3):645-678, 2005.

[16] X. Xu, J. Jager and H-P. Kriegel, A fast parallel clustering algorithm for large spatial databases. *Data Mining and Knowledge Discovery*, 3:263-290 (1999).

[17] L.A. Zadeh. Fuzzy sets. *Inf. Control*, 8:338-353, 1965.

[18] M.L. Zhang, T. Zhang, R. Ramakrishnan, and M. Livny. BIRCH: A new data clustering algorithm and its applications. *Data Mining and Knowledge Discovery*, 1:141-182, 1997.

Chapter 8

Graph-based Clustering

8.1 Introduction

Graph-based clustering or *network clustering* is the process of clustering data or objects represented as a graph. Objects or data points are shown by the vertices of the graph and there is an edge between the objects if they are related according to some metric. In data clustering, we did not require edges between the objects to be clustered whereas in graph clustering, we will specify the neighbors of an entity and these neighbors will have a relation to the object. We will distinguish two cases of clustering where the objects belong to the same cluster if they are closer to each other based on a given objective function in *distance-based clustering*; or *conceptual clustering* where clusters are formed based on the description of the objects. We will assume the former and also the graph representing the network is simple, undirected, weighted or unweighted.

Another distinction is whether the clusters formed overlap or not. In *graph partitioning*, clusters do not overlap, that is, they do not have common members whereas they may overlap in graph clustering. Whether the number of clusters is determined beforehand or not also affects the design of clustering algorithms. If the number of clusters is not known *a priori*, we need to specify an objective function to test and when this function reaches a desired value, the algorithm terminates. There are various parameters called *validation indices* to evaluate the quality (goodness) of a graph-based clustering. One such measure which is also applicable in data clustering is based on the intra-cluster distance which is the average distance of the vertices to the cluster center c_i defined as below :

$$d_{intra} = \frac{1}{n} \sum_{i=1}^{k} \sum_{v \in C_i} d(v, c_i) \tag{8.1}$$

where n is the number of nodes and k is the number of clusters. The *inter-cluster distance* is defined as the minimum distance between the centers of all clusters as $d_{inter} = min(d(c_i, c_j)), 1 \leq i, j \leq k$. The ratio of these two parameters indicates how close the nodes in a cluster are and how far the clusters are from each other [18]. Another measure is based on the density of the graph in which case subgraphs that have a density greater than a given threshold are searched.

Graph partitioning and graph clustering are both NP-hard and therefore, heuristic algorithms are commonly used. They constitute well studied areas of research in computer science and complex networks in general, resulting in numerous algorithms. There is not a single perfect algorithm that suits the needs of all applications but rather, the user needs to select an algorithm for application and possibly tune it for best results. In this chapter, we will first describe graph partitioning and basic algorithms to partition a graph. We will then show the fundamental algorithms to cluster graphs. We will leave the discussion of graph clustering algorithms for biological networks to Chapter 10 when we investigate the PPI networks.

8.2 Graph Partitioning

Formally, *k-way* graph partitioning is the process of dividing the vertices of a graph $G(V, E)$ into k disjoint vertex sets $V_1, ..., V_k$ such that $V_1 \cap V_2 \cap ..V_k = \emptyset$ and $|V_i|$ is approximately n/k. The objective of a partitioning algorithm is therefore to provide partitions of approximately equal number of vertices and the minimum total number of edges among the partitions for unweighted graphs. For a general vertex and edge weighted graph $G(V, E, W_V, W_E)$ where W_V is the set of weights for vertices and W_E is the set of weights for edges of G, a graph partitioning algorithm aims to provide partitions with approximately total equal weights of vertices in each partition and a total minimum weight of edges among the partitions. In the example graph of Figure 8.1, there are four partitions P_1, P_2, P_3 and P_4 with approximately equal sizes. The total number of inter-partition edges is 4 which is much smaller than the total number of edges which is 18.

In Figure 8.2, we have the same graph with vertex and edge weights where weights of vertices are shown inside them and we need to consider these weights while partitioning by keeping the heavy weight edges within the partitions so that total weight of edges among the partitions is minimized. We also aim to balance the total weights of vertices within each cluster. We now have a very different clustering than of Figure 8.1. The *k-way* graph partitioning problem can be solved by *recursive bisection* by first obtaining a 2-way partition of G and then by further dividing each partition using 2-way partitioning. This method which requires $\log k$ steps is used frequently because of its simplicity.

Graph partitioning has many applications such as VLSI design, image processing and parallel processing where nodes represent parallel processes and edges show the communication between them. Each partition in this case is allocated to a number of processors for balancing the load on them and keeping the inter-processor communications minimum eases the communication burden on the processors and hence improves parallelism.

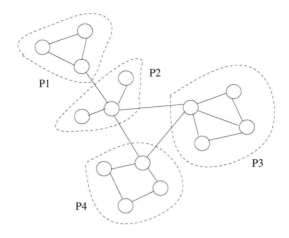

Figure 8.1: Partitions of an unweighted graph

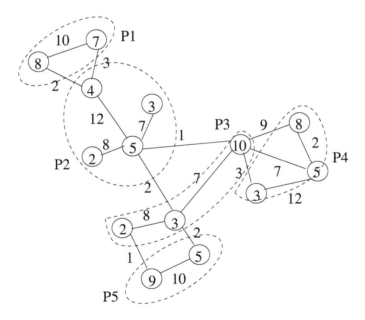

Figure 8.2: Partitions of a weighted graph

8.2.1 *BFS-based Partitioning*

For an unweighted simple graph $G(V,E)$, BFS algorithm can be used to obtain partitions as follows. We first run BFS algorithm of Section 3.4 on a graph $G(V,E)$, starting from the root vertex r to obtain a BFS tree rooted at r. We modify the al-

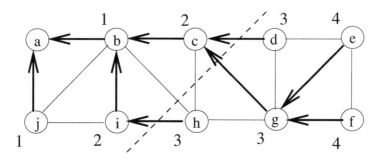

Figure 8.3: *BFS_Part* **algorithm example**

gorithm such that each vertex $v \in V$ has a level l showing its distance to the root as number of hops in the end and this level is recorded. The BFS tree of an example graph is shown in Figure 8.3. There are three types of edges in this graph; BFS *tree edges* shown by bold arrow lines pointing to the parent of a node on the BFS tree, *horizontal edges* that may exist between vertices of the same level and *inter-level edges* which are between adjacent levels but are not part of the BFS tree. For example edge (b,i) belongs to BFS tree; edge (d,g) is a horizontal edge between two nodes of the same level and edge (d,e) is an inter-level edge in this graph. The horizontal and interlevel edges are not part of the BFS tree.

These three types of edges are the only possible edges in the graph, and if we partition this graph into two sets of vertices as V_1 and V_2 such that V_1 consists of vertices that are at level L or lower and V_2 consists of vertices that are at level $L+1$ or higher, the partitioning line will only cross BFS tree or inter-level edges. Therefore, if we choose to cut the graph into two vertex sets such that the number of vertices at level L or lower is approximately equal to the number of vertices at level $L+1$ or higher, we will have achieved the first requirement of the partitioning which is the even distribution of vertices in each partititon. The cut line in Figure 8.3 shows such a partitioning where each partition has four vertices. Since the cut line separates two different levels, it will not cut a horizontal edge, resulting in fewer edges in the cut. For a k-way partitioning, new cut lines may be obtained by recursively cutting the vertices between levels 0 and L and $L+1$ and maximum level.

The BFS algorithm with $O(n+m)$ complexity needs to be executed once but we need to modify this algorithm to record the number of vertices at each level at an array $L[k]$ where k is the maximum level and each entry $L[i]$ has the count of vertices at level i. We can then simply start adding the values of $L[i]$ and when this value is within some offset e from $n/2$, the value of i is recorded. All vertices v_0 (root) and v_i are then in the first partition V_1, and the rest of the vertices are in the partition V_2. For the example of Figure 8.3, $L = \{1,2,2,3,2\}$. Starting with level 0, we add the entries of L until we reach $\lfloor n/2 \rfloor$ which is 5 and the last index that has an accumulated sum of 5 including the root node is 2, therefore, we include all vertices up to and including level 2 in the first partition and the rest in the second partition. Further partitioning of these partitions can be provided by a recursive algorithm (see Ex. 4).

8.2.2 Kernighan-Lin Algorithm

The Kernighan-Lin (KL) algorithm improves the quality of an already existing partition by iteratively swapping vertices between the partitions [18]. It is an early partitioning algorithm dating back to 1970, originally developed to place electronic circuits in printed circuit boards to minimize the connection between the boards. It assumes that the graph is coarsely partitioned by another algorithm beforehand.

Formally, given a graph $G(V, E, w_E)$ with weighted edges, $2n$ vertices, and its two partitions V_1 and V_2, the algorithm finds two partitions A and B of equal size in G with minimum total weight of intersecting edges between them. Furthermore, $A \cap B = \emptyset$, meaning overlaps, are not allowed. This problem is reduced to a max-flow min-cut problem without the first requirement. In other words, the aim of the algorithm is to minimize the total cost function:

$$\sum_{(a,b)\in(A\times B)} w_{ab} \tag{8.2}$$

To accomplish this, it iteratively exchanges two vertices to reduce the cut size. The internal cost of a vertex $a \in A$ is defined as the sum of the weights of edges in A that a is connected as follows:

$$I_a = \sum_{(a,b)\in A} w_{ab} \tag{8.3}$$

and the external cost of a is the sum of the weights of edges with a in one endpoint and a vertex $b \in B$ at the other end point of these edges as follows:

$$E_a = \sum_{a\in A, b\in B, (a,b)\in E} w_{ab} \tag{8.4}$$

The difference between the external cost and the internal cost, $D(a) = E(a) - I(a)$ shows the cost reduction by moving a to b. The gain obtained by swapping vertices a and b has to consider the weight of edge between them as it will contribute twice to the total cost now:

$$g_{ab} = D_a + D_b - 2w_{ab} \tag{8.5}$$

In essence, the KL algorithm swaps a set of vertices between the two partitions that provide the maximum gain at each iteration. The algorithm consists of a number of passes as shown in Alg. 8.1.

In each pass, a pair of vertices a and b whose swapping results in highest gain are locked and are removed from further consideration in the current pass. The D values for the remaining vertices are computed again and the next pair with the highest gain is locked. This process continues until all vertices are locked. At the end of the

---

Algorithm 8.1 *KL_Part*

1: **Input** : $G(V,E), V_A, V_B$ with $|V| = 2n$ ▷ two partitions of G
2: **Output** : A, B such that $|V_A| = |V_B|, V_A \cup V_B = V, V_A \cap V_B = \emptyset$
3: **repeat**
4:      **for all** $v \in V$ **do**   ▷ start a pass
5:          **compute** $D(v)$
6:      **end for**
7:      **for** $i = 1$ to n **do**
8:          **find** unlocked $a \in V_A$ and $b \in V_B$ swapping results in largest gain g_i
9:          **lock** a and b
10:          **for all** $v \in V \setminus \{a,b\}$ **do**
11:              **compute** $D(v)$ as if a and b are swapped
12:          **end for**
13:      **end for**
14:      **find** k where $G_k = \sum_{i=1}^{k} g_i$ is maximum
15:      **if** $G_k > 0$ **then**
16:          **swap** $v_{A1}, ..., v_{Ak}$ with $v_{B1}, ..., v_{Bk}$
17:      **end if**
18:      $\forall v \in V$, **unlock** v
19: **until** $G_k \leq 0$

---

pass, the sequence of the sets of vertices locked which results in the total highest gain is selected and all of the vertices in these sets in the sequence are swapped. The algorithm continues with the passes until the total gain obtained described above becomes negative. Figure 8.4 displays an example weighted graph of six vertices which is initially partitioned to two sets $P_1 = \{a,b,c\}$ and $P_2 = \{d,e,f\}$ with a cut size 13 between them.

The cost reduction for all the vertices in this graph is as follows:

$$D_a = E_a - I_a = 7 + 2 - 3 = 6$$
$$D_b = E_b - I_b = 1 - 3 - 5 = -7$$
$$D_c = E_c - I_c = 3 - 5 = -2$$
$$D_d = E_d - I_d = 3 - 4 - 2 = -3$$
$$D_e = E_e - I_e = 3 - 4 - 3 = -4$$
$$D_f = E_f - I_f = 7 - 2 - 3 = 2$$

We can now calculate the gains for swapping all pairs of vertices:

$$g_{af} = D_a + D_f - 2w_{af} = 6 + 2 - 14 = -6$$
$$g_{ae} = D_a + D_e - 2w_{ae} = 6 - 4 - 4 = -2$$
$$g_{ad} = D_a + D_d - 2w_{ad} = 6 - 3 - 0 = 3$$

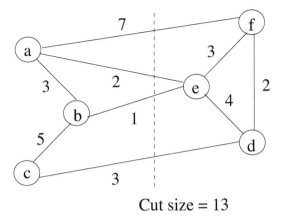

Cut size = 13

Figure 8.4: The initial partition

$$g_{bd} = D_b + D_d - 2w_{bd} = -7 - 3 - 0 = -10$$

$$g_{be} = D_b + D_e - 2w_{be} = -7 - 4 - 2 = -13$$

$$g_{bf} = D_b + D_f - 2w_{bf} = -7 + 2 - 0 = -5$$

$$g_{cf} = D_c + D_f - 2w_{cf} = -2 + 2 - 0 = 0$$

$$g_{cd} = D_c + D_d - 2w_{cd} = -2 - 3 - 6 = -11$$

$$g_{ce} = D_c + D_e - 2w_{ce} = -2 - 4 - 0 = -6$$

The exchange of vertices *a* and *d* results in maximum gain, so we swap them by preserving the topology of the graph. We also lock them so they are not processed again and obtain the graph of Figure 8.5.

We now calculate the new cost reduction values for the unlocked vertices in the new graph as follows:

$$D'_b = E'_b - I'_b = 4 - 5 = -1$$

$$D'_c = E'_c - I'_c = 0 - 8 = -8$$

$$D'_e = E'_e - I'_e = 5 - 5 = 0$$

$$D'_f = E'_f - I'_f = 2 - 10 = -8$$

The new gain values for the unlocked vertices then becomes:

$$g'_{be} = D'_b + D'_e - 2w_{be} = -1 + 0 - 2 = -3$$

$$g'_{bf} = D'_b + D'_f - 2w_{bf} = -1 - 8 - 0 = -9$$

$$g'_{ce} = D'_c + D'_e - 2w_{ce} = -8 + 0 - 0 = -8$$

$$g'_{cf} = D'_c + D'_f - 2w_{cf} = -8 - 8 - 0 = -16$$

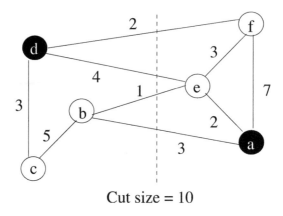

Cut size = 10

Figure 8.5: Vertices *a* and *d* are swapped and locked

Although the maximum gain is now negative, we still exchange vertices *b* and *e* which results in maximum gain, by preserving the topology of the graph. We also lock them so they are not processed again and obtain the graph of Figure 8.6.

The cost function for the only remaining unlocked vertices *c* and *f* in this new graph is then:

$$D_c'' = E_c'' - I_c'' = 5 - 3 = 2$$
$$D_f'' = E_f'' - I_f'' = 5 - 7 = -2$$

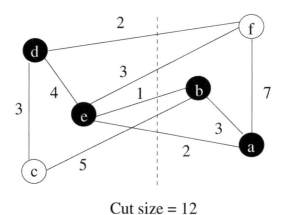

Cut size = 12

Figure 8.6: Vertices *b* and *e* are swapped and locked

Swapping these vertices results in the gain:

$$g''_{cf} = D''_c + D''_f - 2w_{cf} = 2 - 3 - 0 = -1$$

which is the maximum gain as there are no other pairs of unlocked vertices left. We are at the end of the first pass and can now check the accumulated maximum gains obtained at each iteration as follows:

$$g = 3$$
$$g + g' = 3 - 3 = 0$$
$$g + g' + g'' = 3 - 3 - 0 = 0$$

The first gain due to the swapping of vertices a and d only is maximum so we can swap them and conclude the first pass. The KL algorithm has a complexity of $O(n^2)$ time to find the best pair to exchange in the body of for loop and thus $O(n^3)$ total time per pass. The total time for the algorithm is $O(rn^3)$ where r is the number of passes which is usually a small number less than 10. Further improvements to the KL algorithm are possible, for example, to partition a graph into k equal sized sets of vertices. In this case, the graph can be initially partitioned to arbitrary k equal-sized sets and the algorithm may be applied to each pair of subsets. For initial partitions of uneuqal sizes, dummy isolated vertices may be inserted and after applying the KL algorithm, these dummy vertices may be removed. Also, when the size of the graph is large, swapping may be limited to the vertices near the cut as swapping other vertices has less effect on the cut size [13, 15].

8.2.3 Spectral Bisection

We defined the Laplacian matrix L of a graph $G(V, E)$ as the difference of its adjacency matrix and its degree matrix as $L = D - A$ in Section 2.9. The Laplacian L is real and symmetric; therefore its eigenvalues $\lambda_1 \leq \lambda_2 \leq ... \leq \lambda_n$ and its corresponding eigenvectors are real. Furthermore, all eigenvalues of G are nonnegative, hence L is positive semidefinite. The multiplicity of the zero eigenvalues of G shows its number of connected components. The second eigenvalue called the *Fiedler value* and the corresponding eigenvector, the *Fiedler vector* named in honor of Fiedler [9], have an important property and it can be shown that graph G is connected if and only if $\lambda_2 \neq 0$ [20]. This second eigenvalue is also called the *algebraic connectivity* and shows how well G is connected.

The Fiedler vector can be used to partition a graph as shown by Alg. 8.3 where the algorithm first starts by calculating the Laplacian matrix L and the Fiedler vector F. The median value m_F of F is then found and each value of F is compared with the median. The corresponding vertices to smaller values than m_F are stored in V_1, the vertices that have larger Fiedler vector values are stored in V_2 and for equal values, the partition with the less number of vertices is chosen. It can be shown that this method provides balanced partitions of G [20].

Algorithm 8.2 *Spect_Part*

1: **Input** : $A[n,n], D[n,n]$ ▷ adjacency matrix and degree matrix of G
2: **Output** : V_1, V_2 ▷ two balanced partitions of G
3:   $L \leftarrow D - A$
4:   **calculate** Fiedler vector F
5:   **calculate** median m_F of the entries of F
6:   **for** $i = 1$ to n **do**
7:      **if** $F[i] < m_F$ **then**
8:          $V_1 \leftarrow V_1 \cup \{v_i\}$
9:      **else if** $F[i] = m_f$ **then**
10:         **move** v_i to the smaller partition
11:      **else** $V_2 \leftarrow V_2 \cup \{v_i\}$
12:      **end if**
13: **end for**

8.2.4 Multi-level Partitioning

In order to speed up the process of graph partitioning, *multilevel graph partitioning* (MGP) can be used. This method consists of *coarsening, partitioning* and *uncoarsening* phases. In the coarsening phase, a sequence of smaller graphs $G_i(V_i, E_i)$, $i = 1, .., m$ at each step are obtained from the original graph $G_0(V_0, E_0)$. The smallest graph G_m is then partitioned using a suitable partitioning algorithm in the partitioning phase. Finally, the partitioned graph is uncoarsened by applying a local refinement procedure at each step of uncoarsening as shown in Figure 8.7.

The following algorithm shows how to recursively obtain smaller graphs from the original graph G_0. Once the obtained graph has a size smaller than the threshold, the coarsening phase is over and the partitioning is performed at line 4 followed by expansion and refinement procedures at each level of uncoarsening.

Algorithm 8.3 *MGP_Alg*

1:   **procedure** MGP$(G_i(V_i, E_i))$
2:      **if** $|V_i| < threshold$ **then**
3:          **partition** G_i to G_{i0} and G_{i1}
4:          **return** (G_{i0}, G_{i1})
5:      **else**
6:          **coarsen** G_i to G_{i+1}
7:          $G_{(i+1)0}, G_{(i+1)1} \leftarrow MGP(G_{i+1})$
8:          **expand** $G_{(i+1)0}, G_{(i+1)1}$ to G_{i0} and G_{i1}
9:          **improve** partition to obtain $G_{i0'}, G_{i1'}$
10:        **return** $(G_{i0'}, G_{i1'})$
11:      **end if**
12: **end procedure**

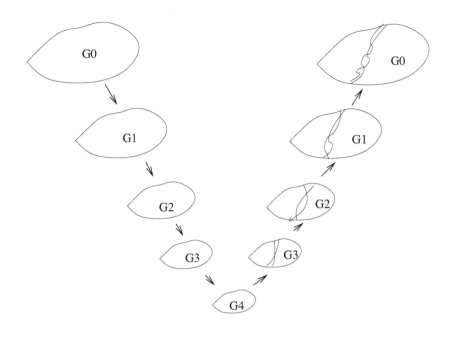

Figure 8.7: Phases of multilevel graph partitioning

We will show a matching-based MGP procedure where a matching M of a graph $G(V,E,W_V,W_E)$ is a subset of its edges such that no vertex in this set shares an endpoint. In this method, a maximal matching MM_i of the graph G_i is obtained at each step i after which *contraction* is performed. The vertices u and v incident to an edge $(u,v) \in MM$ is contracted to a new vertex s such that $w_s = w_u + w_v$, that is, the weights of vertices are added to form the weight of the new vertex. The weights of edges incident to u and v are also added to produce a new edge from the vertex s with this sum.

Two fundamental approaches proposed for matching are *randomized matching* (RM) [13] and *heavy edge matching* (HEM) [14]. Alg. 8.4 shows the pseudocode for obtaining a RM of a graph. It has the same idea of the random matching algorithm of Section 6.4 but applied differently by selecting a random unmatched vertex v rather than a random edge. It then selects a random neighbor u of v that is not connected to any vertices which are the endpoints of edges included in the current matched edges in MM; includes edge (u,v) in the matching and removes both u and v from the graph.

The HEM-based MGP method selects the vertex u that is incident to the heaviest edge of vertex u rather than a random vertex. It is possible to miss a much heavier edge that is incident to the selected vertex u in this method. In another method called the *heaviest weight edge matching* (HWEM), the edges are sorted with respect to their weights and edges are selected from the heaviest to the lighter ones where applicable [13]. Sorting at each step brings considerable computational costs and for this

---

Algorithm 8.4 *RM_Alg*

1: **Input** : $G(V,E)$ ▷ weighted or unweighted
2: $V' \leftarrow V$, $MM \leftarrow \emptyset$
3: **while** $V' \neq \emptyset$ **do**
4:      **select** any $v \in V'$ not connected to any vertex included as end points in MM
5:      **select** any $u \in N(v)$ which is not matched
6:      $MM \leftarrow MM \cup \{(u,v)\}$
7:      $V' \leftarrow V' \setminus \{u,v\}$
8: **end while**

---

reason, this method is favorable only in the later stages of the coarsening when graphs are small. Figure 8.8 displays the implementation of RM, HEM and HWEM heuristics in the same weighted graph where the initial randomly chosen vertices shown in gray are the same for RM and HEM to compare them. We have 82, 77 and 63 for total weights of the cut sets for RM, HEM and HWEM for this particular example. HEM may miss a heavy edge incident to the opposite vertex of the matching of a randomly selected vertex such as the edge with weight 9 in the upper left in this graph.

The terminating condition for the coarsening step is when the graph obtained is small, possibly having less than 100 vertices so that even time-consuming partitioning algorithms can be used or when the obtained smaller graphs do not change significantly. Once the coarsening phase is over, we need to partition the small graph into the required k partitions of approximately equal size. In the uncoarsening phase, the smallest and partitioned graph G_m is projected back to the original graph through a sequence of graphs $G_{m-1}, G_{m-2}, ..., G_0$ by inflating the vertices of G_{i-1}. At each step of uncoarsening, a local refinement procedure such as the Kernighan-Lin algorithm is frequently used.

8.2.5 Parallel Partitioning

Parallel MGP algorithms have been the focus of several studies due to the nature of the problem which allows parallelization in few ways. The two types of parallelism that can be achieved in a k-way MGP algorithm as pointed in [14] are using the recursive structure of the algorithm and in the bisection step. First, a single processor computes two partitions of G and then two processors compute two partitions of these partitions in parallel using a total of $\log p$ processors. The second type of parallelism is achieved when the bisection is performed in parallel. Karypis and Kumar proposed an algorithm that exploits both types of parallelism. Initially, all of the available processors bisect the original graph G into G^0 and G^1 after which half of the computing elements bisect G^0 and the other half bisect G^1 resulting in four subgraphs G^{00}, G^{01}, G^{10}, and G^{11} and so on; then the graph G is divided into p partitions after $\log p$ steps [14]. They showed that the overall run time of the algorithm is $\frac{n \log k}{\sqrt{k}}$ where k is the number of partitions and n is the number of vertices. Different from the scheme outlined, Erciyes et. al. proposed to use fixed centers and grow the partitions around these centers in parallel in [8] and showed that this method provides favorable partitions at the cost of initial placement of centers.

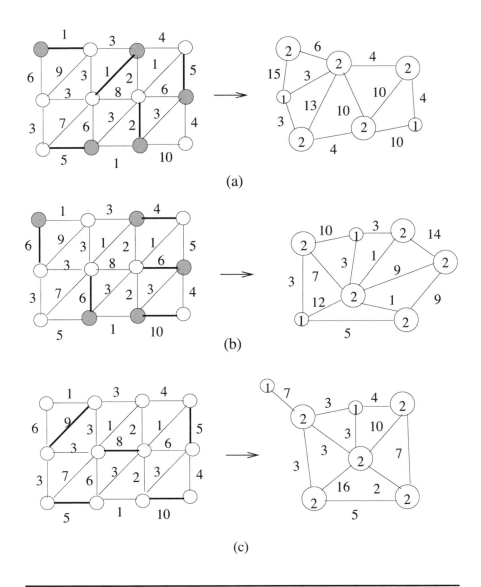

(a)

(b)

(c)

Figure 8.8: a) RM. b) HEM. c) HWEM heuristics to partition a graph

8.3   Graph Clustering

Graph clustering in this context makes use of the neighborhood of the vertices while clustering. Sometimes, this neighborhood can be quantified in terms of distances in which case we try to cluster vertices that are close to each other. In many cases however, neighbor status is what we search when nodes are placed in the same cluster. In this section, we will investigate clustering algorithms that make use of the topological properties of the graph that represents a complex network.

8.3.1 MST-based Clustering

Different MST-based clustering algorithms have different objective functions in general but they all first construct an MST $T(V,E')$ of the graph $G(V,E)$ where $E' \subseteq E$, using Prim's MST algorithm or Kruskal's algorithm we saw in Section 3.5. These algorithms then partition T into k subtrees, k being the number of partitions. Whether k is known beforehand or the partitioning of T is performed until some objective function is met is one major difference between these algorithms.

We will describe a simple algorithm that finds the given k clusters of a graph G using its MST T. This algorithm called *MST_Clust* removes the longest (or heaviest weight) edge from T in the first iteration resulting in a forest F_1 with two subtree components each representing a cluster. It continues to remove the heaviest weight edges from the forests at each iteration for $k-1$ times to find k clusters. Alg. 8.5 shows the pseudocode of *MST_Clust* where *BFS_label* function simply finds connected components starting from one end of the removed edge and includes all of these nodes in the cluster C_i, after removing the heaviest edge from the tree T. Figure 8.9 displays the operation of this algorithm in an example graph.

Algorithm 8.5 *MST_Clust*

1: **Input** : $T(V,E)$ ▷ MST of the graph
2: **Output** : $C_1,...,C_k$ ▷ k clusters
3: **int** $n\_clusts, n\_rows \leftarrow n$
4: **set of points** $C[n] \leftarrow \{\emptyset\}$ ▷ array of clusters
5: $D[n,n] \leftarrow$ distances between points
6: **int** $i \leftarrow 1, S \leftarrow \emptyset$
7: **while** $i \leq k$ **do**
8:      **find** $\{a,b\} \in D$ such that $d(a,b)$ is a maximum real number
9:      $D[a,b] \leftarrow \infty$
10:     $BFS\_label(a,C_i)$
11:     $BFS\_label(b,C_{i+1})$
12:     $i \leftarrow i+1$
13: **end while**

An improvement to this algorithm can be achieved for the cases when k is not known beforehand by using the objective function and a threshold value l_0 for edge weight where any edge $(u,v) \in E'$ that has a value larger than l_0 is removed from T. At each step of the algorithm, the ratio of the intra-cluster distance and the inter-cluster distance is calculated for the newly formed forest and the threshold l_0 is updated. The terminating condition of the algorithm is when the minimum value of

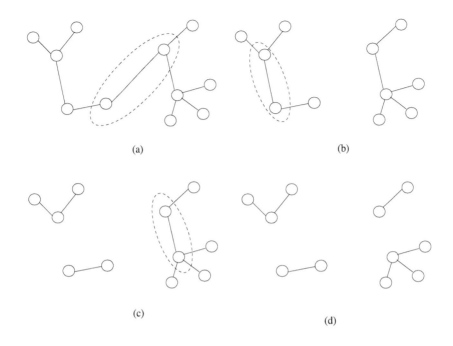

Figure 8.9: *MST_Clust* **example**

the intra-and inter-clustering ratio is found. The vertices belonging to subtrees of the forest for this condition is the output of this algorithm.

Figure 8.9 displays the execution of *MST_Clust* algorithm in an MST of a graph where the length of edges are proportional to their distances. The required number of partitions (k) is four, the algorithm is run three times and the longest edge is removed at each iteration resulting in four clusters shown in (d). MST based clustering is used in various applications including biological networks [22, 21, 23].

8.3.2 Clustering with Clusterheads

A *clusterhead* (CH) of a cluster in a graph is a vertex that is approximately equidistant to all other vertices of the cluster. In a one-hop cluster, all vertices are directly connected to the CH. Figure 8.10 displays three one-hop clusters C_1, C_2 and C_3 with CHs where cluster C_1 is disjoint from the other clusters, and C_2 and C_3 have a common element x.

In a k-hop cluster, any vertex in a cluster is at most k-hops away from the CH. Choosing a clusterhead in this manner eases the formation of the cluster and also a CH can be used as a representative of the cluster to manage various functions of the cluster such as the membership. A dominating set (DS) of a graph $G(V,E)$ was defined in Section 6.3 as the set $D \in V$ such that any vertex not in this set is adjacent to at least one vertex in this set. A connected dominating set (CDS) is a DS with

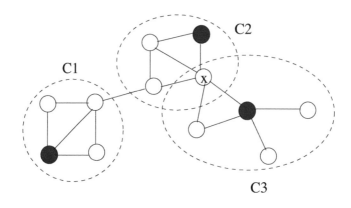

Figure 8.10: One-hop clusters with clusterheads

a path between every pair of vertices in the DS. If we can form a DS of a graph using any of the algorithms described in Section 6.3, we will be able to form one-hop clusters with CHs where any node $v \in$ DS will be a CH and all neighbors of a CH will be the members of the cluster formed by that CH.

The idea of k-hop cluster formation can be combined with the DS formation to yield k-hop dominating sets which may be used to form k-hop clusters. A vertex in a k-hop dominating set is either included in the dominating set or has a distance of at most k hops to a vertex that is in the dominating set. Figure 8.11 shows 2-hop cluster formation using a 2-hop connected dominating sets where three clusters $C_1 = \{a,b,c,d,e,p,r,o,y,z,i,g\}$, $C_2 = \{y,x,d,c,b,r,z,,l,k,i,o,n,j,f,g,h\}$ and $C_3 = \{n,o,i,k,l,q,r,m,y,x,g\}$ are formed around the clusterheads x, y and z respectively. These clusters have a number of common elements; for example, vertices o and r are members of all clusters and vertex g is a member of clusters C_1 and C_2. Clustering using dominating sets is a well studied and researched topic in ad hoc wireless networks as we will describe briefly in Chapter 15.

8.4 Discovery of Dense Subgraphs

A dense subgraph of a graph has an average density which is significantly higher than the density found in the rest of the graph. The dense regions in a graph usually exhibit some kind of activity in these regions when compared with other regions of the graph.

Density can be absolute in which case we evaluate the density of the graph using some absolute metric or relative where we compare the densities of components with respect to each other. In this context, graph clustering is mainly a relative density evaluation method. We will describe algorithms to find dense regions of graphs in absolute terms in this section. Cliques are such absolute measures where the distance between any pair of vertices of a clique is unity and each vertex in an n-clique has

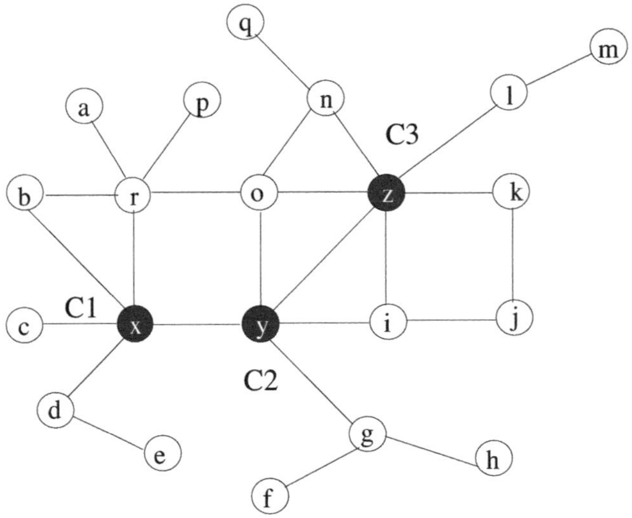

Figure 8.11: Cluster formation using dominating sets

$n-1$ neighbors. Searching for a clique in a graph therefore, does not have any relation to the densities in the rest of the graph. However, cliques may be rare, in which case it is sensible to search for structures which are not exactly cliques but bear some attributes of cliques to some extent.

8.4.1 Definitions

Density of an edge weighted graph $G(V,E,w)$ is the ratio of the total weight of its edges to the number of possible number of edges. Since the total number of possible edges is a graph with n nodes is $n(n-1)/2$, the density of a weighted graph is as follows:

$$den(G) = \frac{2\sum_{(u,v)\in E} w_{(u,v)}}{n(n-1)} \tag{8.6}$$

where $w_{(u,v)}$ is the weight of edge (u,v). If the graph G is unweighted, the density is defined as the ratio of the number of existing edges to the number of possible edges as:

$$den(G) = \frac{2m}{n(n-1)} \tag{8.7}$$

For directed graphs, we need to divide the densities by 2 and a clique has a density of 1. We will now define various dense structures in a graph.

Definition 11 *k*-**clique** *In a k-clique subgraph G' of G, the shortest path between any two vertices in G' is at most k. Paths may consist of vertices and edges external to G'.*

Definition 12 *k*-**club** *In a k-club subgraph G' of G, the shortest path between any two vertices which consists of vertices and edges in G', is at most k.*

Definition 13 **quasi-clique** *A quasi clique is a subgraph G' of G where G' has at least $\gamma|G'||G'| - 1)/2$ edges. In other words, a quasi-clique of size m has γ fraction of the number of edges of the clique of the same size.*

Definition 14 *k*-**core** *In a k-core subgraph G' of G, each vertex is connected to at least k other vertices sin G'. A clique is a (k-1) core.*

Definition 15 *k*-**plex** *A k-plex is a subgraph G' of G, each vertex is connected to at least n-k other vertices in G'. A clique is a 1-plex.*

When searching for dense components of graphs, we should consider whether these components overlap and whether a minimum size for the dense component needs to be specified. Also, whether we should accept all results that meet the criteria or select only the highest ranking components [1].

8.4.2 Clique Algorithms

We will describe two algorithms that find cliques in graphs using enumeration. Both of these algorithms attempt to form a clique by extending a set of vertices until a clique is found. The first algorithm is iterative and the second one uses the branch and bound method.

8.4.2.1 The First Algorithm

We will describe a simple algorithm called *Clique_Clust* that finds the cliques of a graph as described in [7]. Given a graph $G(V, E)$, it starts by first selecting the highest degree vertex $v \in V$ and includes v in the current clique C_i. However, the clique to be discovered may not be of size Δ. It then searches the neighbors of v and if they all have a degree of Δ, they are included in C_i. The search of the elements of the C_i continues until neighbors of the current vertices are investigated. Once C_i is formed, the algorithm selects the highest degree vertex from the remaining vertices of the graph and it terminates when all of the vertices are included in a clique as shown in Alg. 8.6, and the cliques obtained with this algorithm on a sample graph are shown in Figure 8.12. The highest degree vertices are c, d and i, and d is chosen at random initially. The run time for this algorithm is $O(n)$ as each vertex will be considered for a clique in the inner *while* loop.

Table 8.1 shows the iteration of the algorithm. Although the algorithm starts from the highest degree vertex, it may not discover the clique with the highest size first, as shown in this example.

Algorithm 8.6 *Clique1_Alg*

1: **Input** : $G(V,E)$
2: **Output** : $S = \{C_1,...,C_k\}$ ▷ k cliques
3: **int** $i \leftarrow 1, S \leftarrow \emptyset$
4: **while** $V \setminus S \neq \emptyset$ **do**
5:     $C_i \leftarrow \emptyset$
6:     $V' \leftarrow V \setminus S$
7:     **while** $V' \neq \emptyset$ **do**
8:         $v \leftarrow max(deg(u)|u \in V')$ ▷ find the max degree vertex v in V'
9:         $C_i \leftarrow C_i \cup \{v\}$ ▷ include v in current cluster
10:         $V' \leftarrow V' \cap N(v)$ ▷ consider only neighbors of v next
11:     **end while**
12:     $S \leftarrow S \cup \{C_i\}$ ▷ add C_i to clusters
13:     $i \leftarrow i+1$
14: **end while**

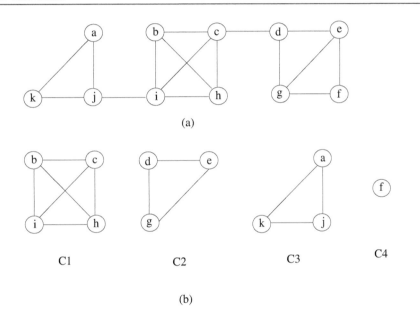

(a)

(b)

Figure 8.12: *Clique_Clust* **example. a) The graph** G. **b) Clusters** C_1, C_2, C_3 **and** C_4

8.4.2.2 The Second Algorithm

An algorithm to find cliques using enumeration was proposed by Bron and Kerbosch [4] where a subset of vertices is extended recursively using the branch and bound method until the clique is maximal. Alg. 8.7 shows the operation of this algorithm where C is the output clique, C_C is the vertex set of the candidate clique and C_N is

Table 8.1: *Clique1_Alg* **Execution**

iterations	1		2	3	4
C_1	$\{c\}$		$\{c,i\}$	$\{c,i,b\}$	$\{c,i,b,h\}$
V'	$\{b,d,i,h\}$		$\{b,h\}$	$\{h\}$	$\{\emptyset\}$
S	$\{\{c,i,b,h\}\}$				
C_2	$\{g\}$		$\{g,e\}$	$\{g,e,d\}$	
V'	$\{d,e,f\}$		$\{d,f\}$	$\{\emptyset\}$	
S	$\{\{c,i,b,h\},\{g,e,d\}\}$				
C_3	$\{k\}$		$\{a,k\}$	$\{j,a,k\}$	
V'	$\{a,j\}$		$\{j\}$	$\{\emptyset\}$	
S	$\{\{c,i,b,h\},\{g,e,d\},\{j,a,k\}\}$				
C_4	$\{f\}$				
V'	$\{\emptyset\}$				
S	$\{\{c,i,b,h\},\{g,e,d\},\{j,a,k\},\{f\}\}$				

the set of vertices to be excluded. Initially C_C includes all of the vertices and $C_N = \emptyset$ and a vertex that is a member of a candidate set but not included in an exclusion set is added to the set of vertices in each step. The time complexity of this algorithm was $O(3.14^n)$ according to the experimental observations of the authors.

Algorithm 8.7 *Clique_2*

1: **Input** : $G(V,E)$
2: **Output** : $S = \{C_1,...,C_k\}$ ▷ k cliques
3: $C_C \leftarrow V$ includes all of the vertices and $C_N \leftarrow \emptyset$
4: *Clique2_Alg* (C, C_C, C_N)
5: **procedure** *Clique2*(C, C_C, C_N)
6:     **if** $C_C = \emptyset \wedge C_N = \emptyset$ **then**
7:         **return** C
8:     **else**
9:         **for all** $v \in C$ **do**
10:            $C_C \leftarrow C_C \setminus \{v\}$
11:            *Clique2*$(C \cup \{v\}, C_C \cap N(v), C_N \cap N(v))$
12:        **end for**
13:    **end if**
14: **end procedure**

8.4.3   *k-core* Algorithm

Finding k-cores of a graph can be performed in polynomial time. The idea of the algorithm proposed by Batagelj and Victorsky is based on the observation that the removal of vertices that have degrees less than k and their incident edges from a graph results in a k-core of the graph [2]. The algorithm starts by inserting vertices in a priority queue based on their degrees, in increasing order. Then at each iteration, the vertex v at the front of the queue is removed, its current degree is assigned as its k-core label and the degrees all of the vertices in the queue that are neighbors of v are decreased by 1, if their current degrees are higher than the degree of v. The procedure is repeated until the queue does not have any more vertices and the k-cores consist of vertices which have at least k labels. The pseudocode of this algorithm is shown in Alg. 8.8.

Algorithm 8.8 $k - core\_Alg$

 1: **Input** : $G(V,E)$
 2: **Output** : k-core values of vertices
 3: $Q \leftarrow$ degree-sorted vertices of G
 4: **while** $Q \neq \emptyset$ **do**
 5:     $v \leftarrow$ front of Q
 6:     $core_v \leftarrow \delta_v$
 7:     **for all** $u \in N(v)$ **do**
 8:         **if** $\delta_u > \delta_v$ **then**
 9:             $\delta_u \leftarrow \delta_u - 1$
10:         **end if**
11:     **end for**
12:     **sort** Q
13: **end while**

Its implementation in the sample graph of Figure 8.13 is shown in Table 8.2. Batagelj et al. showed that the time complexity of this algorithm is $O(max(m,n))$ and since $m \geq n - 1$ in a connected graph, time complexity is $O(m)$.

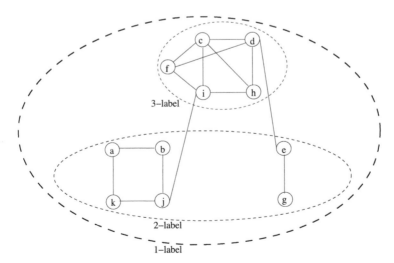

Figure 8.13: Sample graph for finding k-cores 1-core is all nodes; 2-core is 2-label plus 3-label and 3-core is 3-label

Table 8.2: k-core **Algorithm Execution**

iterations	Q	(degrees:	1 to 4)		k=1	k=2	k=3
1	g;	a,b,e,k;	f,h,j;	c,d,i	$\{g\}$	$\{\emptyset\}$	$\{\emptyset\}$
2	e;	a,b,k;	f,h,j;	c,d,i	$\{e,g\}$	$\{\emptyset\}$	$\{\emptyset\}$
3		a,b,k;	d,f,h,j;	c,i	$\{e,g\}$	$\{a\}$	$\{\emptyset\}$
4		b,k;	d,f,h,j;	c,i	$\{e,g\}$	$\{a,b\}$	$\{\emptyset\}$
5		k,j;	d,f,h;	c,i	$\{e,g\}$	$\{a,b,k\}$	$\{\emptyset\}$
6		j;	d,f,h;	c,i	$\{e,g\}$	$\{a,b,k,j\}$	$\{\emptyset\}$
7			f,h,i,d;	c	$\{e,g\}$	$\{a,b,k,j\}$	$\{f\}$
8			h,i,d,c;		$\{e,g\}$	$\{a,b,k,j\}$	$\{f,h\}$
9			i,d,c;		$\{e,g\}$	$\{a,b,k,j\}$	$\{f,h,i\}$
10			d,c;		$\{e,g\}$	$\{a,b,k,j\}$	$\{f,h,i,d\}$
11			c;		$\{e,g\}$	$\{a,b,k,j\}$	$\{f,h,i,d,c\}$

8.5   Chapter Notes

In this chapter, we have reviewed fundamental methods of graph partition, graph clustering and discovering dense regions in graphs. All of these methods have different goals and different procedures to attain these goals. Graph partitioning produces a number of disjoint vertex sets with no common vertices between them but overlapping clusters are allowed in graph clustering. The number of partitions is usually specified beforehand in graph partitioning and recursive algorithms that divide the graph into two partitions at each recursion are commonly used by various algorithms. Multilevel graph partitioning is an effective method consisting of coarsening, partitioning and uncoarsening phases. Some graph clustering methods rely on methods we have seen before such as independent sets and dominating sets. We will further investigate the clustering algorithms for biological networks in Chapter 10 and the community detection algorithms for social networks in Chapter 11, and also look into the clustering algorithms for ad hoc wireless networks in Chapter 13.

Finding optimum graph partitions and clustering are NP-hard, therefore heuristics are widely used. We have seen the random matching and heavy edge matching heuristics for multilevel graph partitioning and also the use of MSTs and dominating sets for graph clustering. Detecting dense subgraphs can be performed by absolute or relative methods. The search of cliques and clique-like structures is the main goal of absolute dense graph detection methods. We have seen algorithms to find cliques and k-cores of graphs. Searching quasi-cliques instead of exact ones provides flexibility in pruning and is pursued in some research studies. For example, the *Quick* algorithm proposed by Liu and Wong uses pruning methods based on the degrees of vertices and DFS to discard the unqualified vertices [17]. The shingling method was first proposed by Broder et al. to find the similarities of Web pages [7]. An algorithm based on shingling is introduced by Gibson et al. to detect dense bipartite subgraphs in large graphs, which transforms dense components of arbitrary sizes to shingles with constant size [25].

In complex networks, our main focus is discovering existing cluster structures in them rather than dividing the graph into predetermined number of balanced partitions; therefore, graph clustering and finding dense components are the common methods pursued by researchers in these networks. Although there are few algorithms to perform graph clustering in parallel such as the parallel MST based clustering in [19], parallel graph clustering for complex networks has received relatively less attention by the researchers and we think this is a potential area of investigation for the researchers of complex networks.

Exercises

1. Describe how to implement a recursive BFS-based graph partitioning algorithm so that a graph G can be divided into k partitions instead of 2.

2. Implement the *BFS_Part* algorithm in the example graph of Figure 8.14 starting from any vertex.

3. The Kernighan-Lin algorithm will be used to refine the two partitions of the graph shown in Figure 8.15. Starting from this initial partition, show a single pass of this algorithm by working out the gain values and locking the associated vertices.

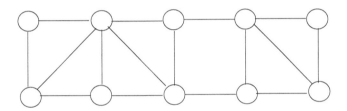

Figure 8.14: Sample graph for Ex. 2

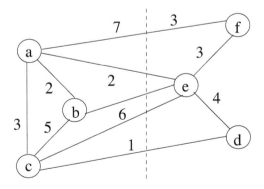

Figure 8.15: Sample graph for Ex. 3

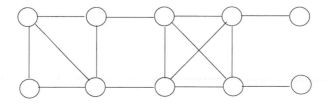

Figure 8.16: Sample graph for Ex. 4

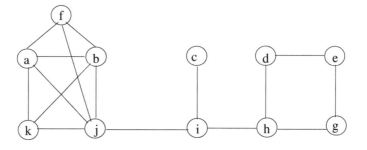

Figure 8.17: Sample graph for Ex. 5

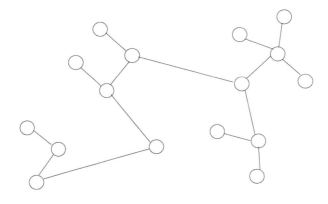

Figure 8.18: Sample graph for Ex. 7

4. Show the execution of *Clique1_Clust* algorithm in the sample graph of Figure 8.16 to find the cliques by showing the contents of each set at each iteration.

5. Show the execution of $k - core$ algorithm in the sample graph of Figure 8.17 to find the k-cores by displaying the iteration steps.

6. Design an algorithm that finds k-hop dominating sets and show the execution of this algorithm in a sample network.

7. Find the four clusters in the MST tree shown in Figure 8.18 where edge lengths are proportional to their weights, by the *MST_Clust* algorithm.

References

[1] C. Aggarwal and H. Wang (Eds.). *Management and Mining Graph Data*, Springer, Chapter 10, 2010.

[2] V. Batagelj and M. Zaversnik. An O(m) algorithm for cores decomposition of networks. CoRR (Computing Research Repository), cs.DS/0310049, 2003.

[3] A.Z. Broder, S.C. Glassman, M.S. Manasse, and G. Zweig. Syntactic clustering of the web. *Comput. Netw. ISDN Syst.*, 29(8-13):1157-1166, 1997.

[4] C. Bron and J. Kerbosch. Algorithm 457: Finding all cliques of an undirected graph. *Commun. ACM*, 16(9):575-577, 1973.

[5] Cormen, Thomas H., Leiserson, C. E., Rivest, R. L. (1990) *Introduction to Algorithms* (1st ed.). MIT Press and McGraw-Hill. ISBN 0-262-03141-8, Section 26.2, 558-565.

[6] P.K. Jana and A. Naik. An efficient minimum spanning tree-based clustering algorithm. In *Proceedings of International Conference on Methods and Models in Computer Science*, 1-5, 2009.

[7] B. H. Junker and F. Schreiber (Eds.). *Analysis of Biological Networks*. Wiley Interscience, Chapter 3, 2008.

[8] K. Erciyes, A. Alp, and G. Marshall. Serial and parallel multilevel graph partitioning using fixed centers. In *Proceedings of SOFSEM*, 127-136, 2005.

[9] M. Fiedler. Laplacian of graphs and algebraic connectivity. *Combinatorics and Graph Theory*, 25:57-70, 1989.

[10] D. Gibson, R. Kumar, and A. Tomkins. Discovering large dense subgraphs in massive graphs. In *Proceedings of 31st ACM Intl. Conf. on Very Large Data Bases*, 721-732, 2005.

[11] A. Gupta. Fast and effective algorithms for graph partitioning and sparse matrix ordering. *IBM Journal of Research and Development*, 41:171-183, 1996.

[12] E. Hartuv and R. Shamir. A clustering algorithm based on graph connectivity. *Information Processing Letters* (76)4:175-181, 2000.

[13] B. Hendrickson and R. Leland. A multilevel algorithm for partitioning graphs. Tech. Rep. SAND93-0074, Sandia National Laboratory, 1993.

[14] G. Karypis and V. Kumar. A parallel algorithm for multilevel graph partitioning and sparse matrix ordering. *Journal of Parallel and Distributed Computing*, 48:71-95, 1998.

[15] G. Karypis and V. Kumar. A fast and high quality multilevel scheme for partitioning irregular graphs. *SIAM J. on Scientific Computing*, 20(1):359-392, 1999.

[16] B. W. Kernighan and S. Lin. An efficient heuristic procedure for partitioning graphs, *The Bell System Technical Journal*, 49(2):291-307, 1970.

[17] G. Liu and L. Wong. Effective pruning techniques for mining quasi-cliques. W. Daelemans, B. Goethals, and K. Morik, Eds., Springer LNAI, ECML/PKDD, 5212:33-49, 2008.

[18] S. Ray and R.H. Turi. Determination of number of clusters in K-means clustering and application in colour image segmentation. In *Proceedings of ICAPRDT99*, pages 137-143, 1999.

[19] V. Olman, F. Mao, H. Wu, and Y. Xu. Parallel clustering algorithm for large data sets with applications in bioinformatics. *IEEE/ACM Trans. Comput. Biol. Bioinform.* 6(2):344-52, 2009.

[20] U. Elsner. Graph Partitioning, A Survey. Technical Report, Technische Universitat Chemnitz, 1997.

[21] Y. Xu, V. Olman, and D. Xu. Clustering gene expression data using a graph-theriotic approach: An application of minimum spanning trees. *Bioinformatics*, 18(2002) 536-545.

[22] C. Zhong, D. Miao, and R.A. Wang. A graph-theoretical clustering method based on two rounds of minimum spanning trees. *Pattern Recognition*, (43)3:752-766, 2010.

[23] W.L. Zhao and Z.G. Zhang. An improved algorithm for clustering gene expression data using minimum spanning trees. *Applied Mechanics and Materials*, 29:2656-2661, 2010.

Chapter 9

Network Motif Discovery

9.1 Introduction

A *network motif* of a graph G is a subgraph of G that appears significantly more than expected in G. Formally, network motif is a subgraph that exists with a much higher frequency in G than expected in a similar random graph to G. A motif of size k is called a k-motif. Milo et. al. were first to discover these over-represented subgraphs in the networks they studied and proposed the network motifs as the basic building blocks of complex networks [20]. Several motifs have been shown to have functional significance in biological networks such as the PPI networks [3] and the gene regulating networks [22] some of which are shown in Figure 9.1.

The feed-forward loop motif for example, was shown to have information filtering capabilities during the gene regulation process. Network motifs may exist in undirected graphs as in the PPI networks or in directed graphs such as the gene-regulating networks. Network motif search is NP-hard, and therefore heuristic algorithms are frequently used. This problem is closely related to the *subgraph isomorphism* where we search the isomorphic occurrences of a small graph in a large graph.

In this chapter, we first describe the network motifs in detail and analyze metrics for their significance. We then look into the related problem of subgraph isomorphism and describe fundamental algorithms that are used for subgraph isomorphism. However, network motif discovery is not only finding an isomorphic subgraph of a graph as we need to asses the number of occurrences of this subgraph to consider it as a motif. In other words, we need to find all occurrences of the motif. Finding all existing motifs of a given size is called *census* and we investigate exact and approximate census algorithms in Section 1.4.

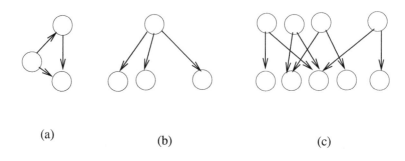

(b)

(c)

Figure 9.1: Motifs in biological networks. a) Feed-forward loop. b) A single input motif. c) A multi-input motif

9.2   Network Motifs

A motif is a connected and usually a small graph m with k number of vertices called its size. A *match* G' of a motif in the target graph G is a subgraph of G which is isomorphic to m. Two possible undirected motifs of size three are shown in Figure 9.2 and Figure 9.3 displays all possible thirteen directed motifs with the size 3. Network motif search and analysis in general are concerned with the more difficult problem of directed networks.

The number of occurrences of a motif m in the graph G is called its *frequency* and there are three variations of this frequency as F_1, F_2 and F_3 in G [27]. The F_1 frequency is related to all matchings, including overlapping ones of m in G; F_2 is the frequency of edge disjoint matchings, and F_3 is the frequency of edge and vertex disjoint matchings. Figure 9.4 displays the feed-forward loop motifs in a small network. There are four occurrences of this motif and subgraphs m_1 and m_4 are vertex and edge disjoint whereas m_1, m_2, m_3 share vertex a; m_2, m_1, m_2 share vertex f; and m_2, m_3 and m_4 share vertex e. Subgraphs m_1 and m_2 share the edge (a, f) and m_2 and m_3 share the edge (a, e). Examining Figure 9.4, we find F_1 as 4 which includes all matchings m_1, m_2, m_3, m_4; F_2 as 3 for matchings m_1, m_3, m_4, and F_3 as 2 for matchings m_1 and m_4.

Finding the maximum number of unique, nonoverlapping matches of a motif (F_3) is equivalent to the maximum independent set problem we have seen in Section 6.2.

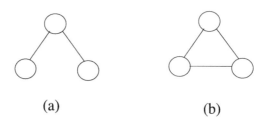

(a)

(b)

Figure 9.2: Undirected network motifs of size 3. a) A triad. b) A triangle

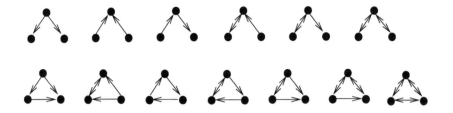

Figure 9.3: All thirteen directed network motifs of size 3

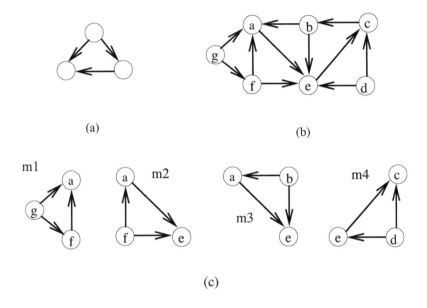

(a)

(b)

(c)

Figure 9.4: Motifs in a graph. a) The feed-forward loop motif. b) The target graph. c) Motifs m_1, m_2, m_3 and m_4 discovered

As this problem is NP-hard [6], finding F_3 of a motif m in a graph G is also NP-hard which necessitates the use of heuristics to compute a lower bound for F_3. Table 9.1 shows the number of possible motifs of sizes 3,...,10 in undirected and directed graphs. Clearly, the number of directed graphs grows much faster than the number of undirected graphs for the same number of vertices.

Table 9.1: Number of Possible Network Motifs

Motif size	3	4	5	6	7	8	9	10
Undirected subgraphs	2	6	21	112	853	$\approx 10^4$	$\approx 10^5$	$\approx 10^7$
Directed subgraphs	13	199	9364	10^6	$\approx 10^9$	$\approx 10^{12}$	$\approx 10^{16}$	$\approx 10^{20}$

9.2.1 Measures of Motif Significance

We need to evaluate the occurrences of the motif m in the target graph and also in random networks to assess its significance. The randomized version R of a graph G which is similar in structure to G is called a *null model*. There will be a number of random graphs $\Psi(G)$ which have similar properties to G, such as the same size of vertices, the same degree sequence etc. We choose a predefined number n of random graphs $\Psi'(G)$ uniformly from $\Psi(G)$ and compute the frequency of a subgraph m in both G and $\Psi'(G)$. If the frequency of m in G is significantly higher than its average frequency in the random graphs of $\Psi'(G)$, m can be considered a network motif. The statistical significance of a motif m in a graph G can be determined by calculating the metrics Z-score, P-value and the motif significance profile using the frequency F_1 as described below:

■ **Z-score**: Z-score of a motif m, $Z(m)$, is the ratio of the difference between the frequency F_1 of m in the target network and its average frequency $\overline{F_{1,r}}$ in a set of randomized networks, to the standard deviation σ_r of the F_1 values in the set of randomized networks as follows:

$$Z(m) = \frac{F_1(m) - \overline{F_{1,r}(m)}}{\sigma_r(m)} \tag{9.1}$$

A motif m with a value of $Z(m)$ that is greater than 2.0 is considered significant [9].

■ **P-value**: The P-value of a motif m, $P(m)$, is defined as the probability that the frequency of the motif m in a randomized network is equal to or greater than the frequency of m in the target network [28]. It is calculated by counting the number of random networks in which the motif occurs more often than the target network and dividing this number by the total number of random networks as follows:

$$P(m) = \frac{1}{n} \sum_{i=1}^{n} \sigma_{R_i}(m) \tag{9.2}$$

where $\sigma_{R_i}(m)$ is a binary variable which is set to 1 if the occurrence of motif m in the random network R_i is equal to or greater than its occurrence in the target graph G. P-values are between 0 and 1, and a smaller P-value indicates more significant motifs; meaning the incidence of m in G is not incidental.

■ **Motif significance profile**: The motif significance profile (SP) consists of a vector of Z-scores of a set of motifs $m_1, m_2, ..., m_k$ normalized to unity as shown in Eqn. (9.3). SP is used to compare different networks with respect to the motifs they contain.

$$SP(m_i) = \frac{Z(m_i)}{\sqrt{\sum_{i=1}^{n} Z(m_i)^2}} \tag{9.3}$$

9.2.2 Generating Null Models

The simplest null model is formed by generating a graph G' which has the same size but different topology with the target graph G. The graph G' can be obtained by randomly rewiring the edges of G, similar to the Watts-Strogatz model process discussed in Section 3.8. However, preserving the degree sequence of G, and the in-degrees and the out-degrees of the nodes in G while generating the random graphs is more commonly used as the topological structure of G is better reserved in this manner. In order to provide such random graphs, two directed edges (a,b) and (c,d) of G are chosen at random, and vertex a is connected to vertex d and c is connected to b by deleting the edges (a,b) and (c,d). This process is known as the Markov chain method and is shown in Alg. 9.1. We need to make sure that the rewired edges do not exist in the graph before forming the new edges.

Algorithm 9.1 *Markov − Chain_Alg*

1: **Input** : $G(V,E)$
2: **Output** : Random graph $G'(V',E')$ similar to G
3:   $G' \leftarrow G$
4: **while** G' is not as randomized as required **do**
5:     **pick** two random edges $(a,b),(c,d) \in E'$
6:     **if** $(a,d) \notin E' \wedge (c,b) \notin E'$ **then**
7:         $E' \leftarrow E' \setminus \{(a,b),(c,d)\}$
8:         $E' \leftarrow E' \cup \{(a,d),(c,b)\}$
9:     **end if**
10: **end while**

The implementation of this algorithm is illustrated in Figure 9.5. This process is repeated for many times and in the end, the in-degrees and out-degrees of all nodes in G are preserved but we have a graph that is very differently connected than G.

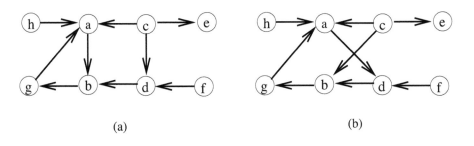

(a)   (b)

Figure 9.5: Generation of a random network. a) The original graph. b) The edges (a,b) **and** (c,d) **are modified**

9.2.3 Hardness of Motif Discovery

There are a number of problems involved in the search of motifs. Firstly, isomorphic motifs have to be grouped together as they are the occurrences of the same motif. This process is equivalent to finding the isomorphic subgraphs of a graph which is a computationally hard problem as we will see in the next section. Secondly, the number of motifs grows exponentially with the size of the motif. For directed graphs, there are clearly many more combinations than the undirected motifs of the same size. We have seen that the number of non-isomorphic motifs of size 3 is two whereas for directed motifs of size 3, there are thirteen combinations. This means searching for motifs which have sizes larger than 5 for example, is a computationally difficult task. We also need to count the occurrences of the isomorphic motifs in the target graph. Finally, we need to asses the statistical significance of the number of motifs found which can be accomplished by generating many random graphs with similar properties with the target graph and evaluating the frequencies of the motifs of the given size in these random graphs to compare these with the target graph. The three metrics defined as Z-score, P-value and SP value to quantify the significance of motif frequencies in these graphs are frequently used.

In the rest of this chapter, we will first investigate the closely related problem of subgraph isomorphism to motif discovery. We will then briefly survey contemporary algorithms and software tools designed for motif discovery.

9.3 Subgraph Isomorphism

Graph isomorphism can be informally stated as checking to see if two graphs that appear different are in fact the same. It is one of the rare problems which is in NP but not known to be in P or NP-complete. Formally, two graphs $G_1(V_1, E_1)$ and $G_2(V_2, E_2)$ are isomorphic, shown by $G_1 \cong G_2$, if there is a bijection $\varphi : V_1 \rightarrow V_2$, such that:

$$\forall v_i, v_j \in V_1, (v_i, v_j) \in E_1 \Longleftrightarrow (\varphi(v_i), \varphi(v_j)) \in E_2 \tag{9.4}$$

This one-to-one mapping is called an *isomorphism* between G_1 and G_2. A graph $G_1(V_1, E_1)$ is isomorphic to a subgraph of a graph $G_2(V_2, E_2)$ if $\exists G_2' \subset G_2$ where $G_1 \cong G_2'$. The bijection in this case is called a *subisomorphism* between G_1 and G_2. Informally, subgraph isomorphism is the search of a smaller graph within a larger graph and it is an NP-complete problem [6]. An *automorphism* of a graph G is an isomorphism with itself. A subgraph isomorphism between two graphs G_1 and G_2 is depicted in Figure 9.6. The subgraph isomorphism has many applications including image processing, pattern recognition, information retrieval and bioinformatics.

9.3.1 Vertex Invariants

Given two graphs $G_1(V_1, E_1)$ and $G_2(V_2, E_2)$, *vertex invariant* is a label $l(v_1)$ assigned to a vertex $v_1 \in G_1$ such that if there exists an isomorphism φ_{12} that maps v_1 to $v_2 \in G_2$, then $l(v_1) = l(v_2)$ [5]. A very common vertex invariant is the degree of

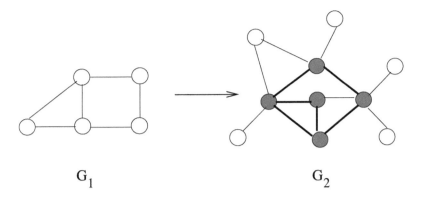

G_1 G_2

Figure 9.6: Subgraph isomorphism

---

a vertex. The isomorphism that maps vertices of G_1 to the vertices of G_2 should preserve the degrees of the vertices that are mapped. However, vertices that have the same degrees in G_1 and G_2 are not necessarily mapped under an isomorphism. The following vertex invariants are used in an isomorphism detection program called *nauty* [15, 5]:

■ **two paths**: The number of neighbors a vertex v has in its two-hop neighborhood.

■ **adjacency triangles**: Sizes of the common neighbors of the neighbors of v that are adjacent.

■ **k-cliques**: Number of different cliques of size k that include v.

■ **independent k-sets**: Number of different independent sets of size k that include v.

■ **distances**: The number of vertices at distances $1,...,n$ from v.

Most of the time, vertex invariants are a combination of these invariants and are calculated using the degree invariant as the base and appending other invariants to this base. Vertex invariants can be calculated in polynomial time; however, direct solution to the isomorphism problem may provide more efficient solutions, as described next.

9.3.2 Algorithms

As an attempt to discover isomorphism between two graphs G_1 and G_2, we define $G_1(k)$ as the subgraph of G_1 induced on the vertices $1,...,k$ and start building an isomorphism by extending the isomorphism to $G_1(k+1)$ by adding a vertex, which is a common method pursued in various research studies [5]. We will now describe two previously designed algorithms that are frequently used and two more recent algorithms for subgraph/graph isomorphism.

9.3.2.1 Ullman's Algorithm

The algorithm proposed by Ullman dates back to 1976 and forms the basis of various subgraph/graph isomorphism algorithms [29]. It uses the following relation between the two isomorphic graphs G_1 and G_2 of sizes n with respective adjacency matrices $A_1[n,n]$ and $A_2[n,n]$:

$$A_2 = PA_1P^T \tag{9.5}$$

where P is a permutation matrix. Existence of such a P matrix indicates that the two graphs are isomorphic. A permutation matrix is an $n \times n$ matrix obtained by the permutation of the rows of an identity matrix of the same size. It therefore has only 0 or 1 elements and there is exactly a single 1 in each row or column. Multiplication of a permutation matrix with its transpose gives the identity matrix as $PP^T = I$. Left multiplication of a matrix A by a permutation matrix rearranges the corresponding columns of A as shown below:

$$\begin{bmatrix} a & b & c \\ d & e & f \\ g & h & i \end{bmatrix} \begin{bmatrix} 0 & 1 & 0 \\ 0 & 0 & 1 \\ 1 & 0 & 0 \end{bmatrix} = \begin{bmatrix} c & a & b \\ f & d & e \\ i & g & h \end{bmatrix} \tag{9.6}$$

and right multiplication by a permutation matrix changes the corresponding rows of A as follows:

$$\begin{bmatrix} 0 & 0 & 1 \\ 1 & 0 & 0 \\ 0 & 1 & 0 \end{bmatrix} \begin{bmatrix} a & b & c \\ d & e & f \\ g & h & i \end{bmatrix} = \begin{bmatrix} g & h & i \\ a & b & c \\ d & e & f \end{bmatrix} \tag{9.7}$$

Multiplication of a matrix from left and right by a permutation matrix changes its rows and columns. Let us assume we need to test isomorphism of the two graphs G_1 and G_2 shown in Figure 9.7. It can be seen that these graphs are isomorphic and vertices are mapped as follows: $a \rightarrow x$, $b \rightarrow y$, $c \rightarrow z$ and $d \rightarrow w$.

We can write the adjacency matrices A_1 and A_2 corresponding to G_1 and G_2 as follows:

$$A_1 = \begin{bmatrix} 0 & 1 & 0 & 1 \\ 1 & 0 & 1 & 1 \\ 0 & 1 & 0 & 1 \\ 1 & 1 & 1 & 0 \end{bmatrix}, \quad A_2 = \begin{bmatrix} 0 & 1 & 1 & 0 \\ 1 & 0 & 1 & 1 \\ 1 & 1 & 0 & 1 \\ 0 & 1 & 1 & 0 \end{bmatrix} \tag{9.8}$$

A close inspection of these two matrices yields the existence of a permutation matrix P that satisfies Eqn. (9.5) to provide the transformation $A_2 = P^T A_1 P$ shown below:

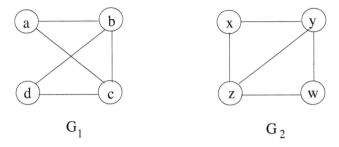

Figure 9.7: Two isomorphic graphs

$$
\begin{bmatrix} 0 & 1 & 1 & 0 \\ 1 & 0 & 1 & 1 \\ 1 & 1 & 0 & 1 \\ 0 & 1 & 1 & 0 \end{bmatrix} = \begin{bmatrix} 0 & 0 & 1 & 0 \\ 0 & 0 & 0 & 1 \\ 0 & 1 & 0 & 0 \\ 1 & 0 & 0 & 0 \end{bmatrix} \begin{bmatrix} 0 & 1 & 0 & 1 \\ 1 & 0 & 1 & 1 \\ 0 & 1 & 0 & 1 \\ 1 & 1 & 1 & 0 \end{bmatrix} \begin{bmatrix} 0 & 0 & 0 & 1 \\ 0 & 0 & 1 & 0 \\ 1 & 0 & 0 & 0 \\ 0 & 1 & 0 & 0 \end{bmatrix}
$$
(9.9)

Ullman's algorithm is based on this concept where a permutation matrix of a partition of the adjacency matrix of the larger graph G_2 is progressively generated up to the size of the small graph and Eqn. (9.5) is checked at each step. Let $S_{k,l}(A_1[n,n])$ be the matrix obtained from A_1 by deleting the rows $k+1,..,n$ and columns $l+1,...,n$ from A_1. Now, for the subgraph isomorphism testing, this equation can be written as [29, 16]:

$$A_1 = S_{l,l}(PA_2P^T) \tag{9.10}$$

where $S_{t,n}(P)$ is an $t \times n$ permutation matrix showing a partial matching from the first t vertices of G_2 onto a subset of vertices of G_1. If there is a permutation matrix P such that $S_{t,t}(A_2) = S_{t,n}(P)A_1(S_{t,n}(P))^T$ then $S_{t,n}$ is an $t \times n$ permutation matrix representing a subgraph isomorphism consisting of vertices $1,..,t$ of G_1 to a subgraph of G_2.

In this algorithm, permutation matrices are generated for the subgraphs of the target graph G_2 and these are then compared with the subgraph G_1 to check isomorphism. An entry in the permutation matrix is set to 1 and a check is done to find if $S_{1,n}$ is a partial matching in which case the procedure *BackTrack* is called which increases the size of the matching up to k. If the size of the partial matching reaches k, a subgraph isomorphism is detected and the permutation matrix P representing the transformation is returned. Alg. 9.2 displays the operation of Ullman's algorithm where the adjacency matrices of the two graphs G_1 and G_2 are processed. The output from the algorithm is the permutation matrix P of the match found. The *BackTrack* procedure is called recursively and if the equality in line 17 is valid up to and including the size n_2 of a subgraph G_1, the permutation matrix P is returned and the occurrence of this subgraph is detected.

Algorithm 9.2 *Ullman_Alg*

1: **Input** : $G_1(V_1,E_1)$, $G_2(V_2,E_2)$ ▷ weighted or unweighted
2:       $P[n_2,n_2]$: a permutation matrix
3: **Output** : $G' \subset G_2$ such that $G' \cong G_1$ if exists
4: $n_1 \leftarrow |V_1|$; $n_2 \leftarrow |V_2|$
5: *BackTrack* $(A_1,A_2,P,1)$
6:
7: **procedure** BACKTRACK(A_1,A_2,P,k)
8:     **if** $k > n_1$ **then**
9:        P represents a subgraph isomorphism from G_1 to G_2
10:        **return**(P)
11:     **end if**
12:     **for** $i = 1$ to n_2 **do**
13:        $p_{ki} \leftarrow 1$
14:        **for all** $j = 1$ to n_1, $j = i$ **do**
15:           $p_{kj} \leftarrow 0$
16:        **end for**
17:        **if** $S_{k,k}(A_1) = S_{k,n}(P)A_2(S_{k,n}(P)^T)$ **then**
18:           *BackTrack* $(A_1,A_2,P,k+1)$
19:        **end if**
20:     **end for**
21: **end procedure**

Ullman's algorithm can be used for graph or subgraph isomorphism and its time complexity is $O(m^n n^2)$ and its space complexity is $O(n^2 m)$ [29, 16] where m and n are the orders of the subgraph and the target graph, respectively.

9.3.2.2   Nauty Algorithm

The *Nauty* algorithm proposed by McCay is a powerful algorithm that is still used to find graph and subgraph isomorphisms [14]. It transforms the graphs to be matched to a canonical form before testing for isomorphism and uses vertex invariants and group theory while searching for graph/subgraph isomorphism. We will briefly describe its main operation using high level procedures as in [5]. A partition $\mathcal{P}$ divides the vertices of a graph G into a number of disjoint sets $V_1, ..., V_m$. A partition that has only singleton sets is called a *leaf partition*. Nauty forms an initial partition by computing the invariants over the whole graph and then invariants are computed for individual partitions to distinguish them. It uses refining a partition and generating the children of a partition as the two main methods of the algorithm. The main operation of Nauty is a depth-first-search of the space partitions where each partition is refined before expanding its children as shown in Alg. 9.3 [5].

Assuming $d(v,Q)$ is the number of neighbors of v in set Q, and the initial partition $\mathcal{P}$ is $\{V_1, ..., V_m\}$ where $\forall a, b \in V_i$, $d(a, V_i) = d(b, V_i)$. A partition is then refined as follows [5]:

Algorithm 9.3 *Nauty_Alg*

```
 1: Input : G₁(V₁,E₁)                          ▷ weighted or unweighted
 2: Output : G₂(V₂,E₂) : a canonical graph
 3:
 4:   P ← partition of a single cell V
 5:   S ← stack containing P
 6: while S ≠ ∅ do
 7:     u ← pop(S)
 8:     if u = leaf partition then
 9:         update(G₂,u)
10:     else
11:         refine(u)
12:         append children of u to S
13:     end if
14: end while
```

---

1. Find $V_i \in P$ that has more than one element.

2. $\forall v \in V_i$, compute $s_v = (d(v,V_1),...,d(v,V_m))$.

3. Divide V_i into subsets such that all vertices in a subset have the same value of s_v.

The procedure is repeated for all V_i including the ones formed after the split, until sets cannot be divided anymore. The children of a partition are generated by selecting the first set V_i that has more than one element and for each vertex $v \in V_i$, a child partition with elements $V_1,..,V_{i-1},\{v\},V_i/\{v\},V_{i+1},..,V_m$ is formed. In other words, a new child of vertex v is formed by dividing the set V_i into two sets where one set contains only v and the other has the elements of V_i excluding v. Nauty checks automorphism of a graph and computes a canonical label which is the adjacency matrix of the smallest automorphism. It uses vertex variants in the refinement process rather than extending automorphism. A detailed description of *Nauty* can be found in [14] and in [5].

9.3.2.3 VF2 Algorithm

Cordella et al. introduced the VF2 algorithm [4] for exact matching in graphs, which is an improvement over their first algorithm in [3]. This algorithm is based on depth-first search and uses a refinement procedure to reduce the search space. A *mapping* M between the two graphs $G_1(V_1,E_1)$ and $G_2(V_2,E_2)$ is defined as a set of vertex pairs (u,v) with $u \in V_1$ and $v \in V_2$. A mapping M is denoted as isomorphism if and only if it is a bijective function that preserves the branch structure of G_1 and G_2 [4]. The mapping at state s is denoted by M_s in state space representation where s is the state representing current set of associations between vertices in G_1 and G_2. The pseudocode for the VF2 algorithm is shown in Alg. 9.4.

Algorithm 9.4 $VF2\_Alg$

1: **procedure** MATCH(G_1, G_2, s)
2:      **Input** : two graphs G_1 and G_2, an intermediate state s
3:      initial state s_0 has $M(s_0) = \emptyset$
4:      **Output** : the mappings between G_1 and G_2
5:      **if** M_s covers all nodes of G_1 **then**
6:          **return**(M_s)
7:      **else**
8:          **compute** $P(s)$ of candidate pairs to be included in $M(s)$
9:          **for all** $p \in P(s)$ **do**
10:             **if** p is compatible with $M(s)$ **then**
11:                 $s' \leftarrow p \cup M(s)$
12:                 **compute** state s' obtained by adding p to $M(s)$
13:                 $Match(G_1, G_2, s')$
14:             **end if**
15:         **end for**
16:         **restore** data structures
17:     **end if**
18: **end procedure**

The mapping function does not have any component in the initial state s_0. If the partial mapping obtained covers all of the vertices of G_1, an isomorphism is found, otherwise, the search tree is enlarged by DFS and a new pair of vertices p is tested to be added to the current state.

The VF2 algorithm introduces a set of rules and the candidate pair of vertices p is tested by a number of pruning rules where p is found compatible with the existing subgraph if it passes all of these tests. Furthermore, whether a consistent state has consistent successor states in k steps is also checked by the *k-look-ahead rules*. The time complexity of this algorithm is $\Theta(n!n)$ in the worst case and $\Theta(n^2)$ in the best case. Space complexity in both cases is $\Theta(n)$ [4].

9.3.2.4 *BM1 Algorithm*

The BM1 algorithm proposed by Batitti and Mascia uses counting of the paths of various types in the subgraph G_1 and the target graph G_2. The main idea of the algorithm is that if there is a path of length d emanating from a vertex v_1 in G_1, the same path originating from v_2 in G_2 should also exist so that (v_1, v_2) pair is considered as part of the matching. In this so called *local-path-based* pruning, paths are counted in the undirected graph corresponding to the original graph where an edge (u, v) is present in the undirected graph if and only if edges (u, v) and (v, u) or both exist in the original graph.

The pre-processing phase of the algorithm finds the number of paths of length up to d of all paths originating from the vertices of G_1 and G_2. A pair of vertices v_1 and v_2 are included in the mapping if the number of paths originating at v_1 and v_2 is compatible. This pruning method can be applied before the VF2 algorithm to further reduce the size of the search tree. The pseudocode for the BM1 algorithm is shown in Alg. 9.5 [2]. The value of the parameter d should be carefully selected as larger d values will result in more vertices pruned at the cost of increased computation to detect them.

Algorithm 9.5 *BM1_Alg*

1: **procedure** COMPATIBLEPATHS(v_1, v_2, d)
2:     **Output** : the mappings between G_1 and G_2
3:     **for all** x **in** $1, ..., d$ **do**
4:         **for all** y **in** $1, ..., 2^d$ **do**
5:             **if** $PathsDS[v_1][x][y] > PathsDS[v_2][x][y]$ **then**
6:                 **return** *false*
7:             **end if**
8:         **end for**
9:     **end for**
10: **end procedure**

Time complexity of the BM1 algorithm is $O(n^d)$ and the memory space required is $n(2^{d+1} - 2)$ [2]. Battiti and Mascia compared BM1 and VF2 and showed that BM1 performs better than VF2 experimentally for randomly generated graphs of various sizes.

9.4 Motif Discovery Algorithms

Two fundamental methods of identifying network motifs are *exact counting* and *subgraph sampling*. Exact counting methods attempt to find every occurrences of a motif whereas only sampled subgraphs are tested in the latter. Finding network motifs consists of three subtasks:

1. Discovery of motifs $m_1, ..., m_n$ which occur more than expected in the target graph G.

2. Grouping of these motifs into isomorphic classes $C_1, ..., C_k$ such that all motifs in a class are topologically equivalent.

3. Determination of these classes that appear at a much higher frequency in G than in random graphs which are topologically similar to G.

The first task requires explicit enumeration of all subgraphs of a given size k and usually has a high time complexity. Many of the discovered motifs may in fact be the same subgraph, we therefore need to classify them into specific classes in the second step, where motifs in the same class are isomorphic to each other. For the third task, we need to generate random graphs that are topologically similar to the target graph G, using the Markov chain or a similar algorithm.

Certain properties of a motif search algorithm affect its performance and accuracy. One such feature is the possibility of *sampling* where only a representative sample of the target graph is searched and therefore an approximate result is obtained. In *biased sampling*, the algorithm is biased towards a particular form of samples; thus, we would need to correct the bias at the end of the algorithm execution. The algorithm may result in finding the same subgraph multiple times using different search paths. In *symmetry breaking* algorithms, a unique subgraph is discovered exactly once. The size of the motif that an algorithm can discover in reasonable time is an indication of its performance. Motif sizes up to 5 are usually considered small, sizes between 6 and 8 are considered medium and any size greater than 8 is large [24].

Motif search algorithms can be broadly classified as *exact census algorithms* where each occurrence of the motifs is recorded; or *approximate census algorithms* where we sample the target graph and search for motifs in the sample graph.

9.4.1 Exact Census Algorithms

Exact census algorithms rely on full enumeration, that is, finding each subgraph of the required size one by one. We will now describe significant exact motif search algorithms.

9.4.1.1 M finder Algorithm

M finder is the first motif finding tool and implements two kinds of motif finding methods as a full enumeration and subgraph sampling [20, 19, 21]. A subgraph in *mfinder* is defined as a set of vertices $V_s \in V$ of a graph $G(V,E)$ with the edges that exist between them in the original graph. In other words, the subgraph G_s obtained is an induced subgraph of G. In the full enumeration method, an edge (u,v) is chosen and a subgraph G' is constructed around this edge. A vertex that is neighbor to any of the vertices of the partial graph G' is added to this graph. When the size of G' is equal to the required motif size, the identity of the isomorphism class is calculated and a hash table containing the frequencies is updated. The set of vertices of G' found are inserted in a hash table to prevent them from being revisited. This process is repeated for all edges of the graph as shown in Alg. 9.6 as adapted from [23].

The operation of *mfinder* is illustrated in Figure 9.8 for a simple undirected graph G with five vertices labeled 1,...,5 and we search motifs of size 3, that is, triads and triangles in this graph. The algorithm returns seven triads and one triangle as shown.

Although each unique subgraph found is inserted in a hash table, the same subgraph may be found multiple times due to symmetries and a further check is needed to verify that the newly found motif is indeed an unsearched one. Wernicke showed that the *mfinder* sampling algorithm may result in undersampling of some motifs and oversampling of some other motifs [31]. Due to the possible multiple searches of the same motif, the memory space and the computational power required by this algorithm are high, making it unsuitable for motifs of large sizes, typically sizes of 5 or more.

Algorithm 9.6 *mfinder*

1:	**Input** : $G(V,E)$ ▷ weighted or unweighted
2:	**int** $2 \le k \le n$
3:	**Output** : Subgraphs of size k
4:	**for all** $(u,v) \in E$ **do** ▷ do it for all edges
5:	$Extend(\{u,v\})$
6:	**end for**
7:	**procedure** EXTEND (G')
8:	**if** $|G'| = k$ **then** ▷ if required size is reached
9:	**if** G' is unique **then**
10:	**increment** frequency of isomorphic class of G'
11:	**else**
12:	**insert** G' in Hash
13:	**for all** $u \in G'$ **do**
14:	**for all** $(u,w) \in E$ **do** ▷ check neighbors
15:	**if** $w \notin G'$ **then**
16:	**if** $G' \cup \{w\}$ is not in Hash **then**
17:	$Extend(G' \cup \{w\})$
18:	**end if**
19:	**end if**
20:	**end for**
21:	**end for**
22:	**end if**
23:	**end if**
24:	**end procedure**

9.4.1.2   *Enumerate Subgraphs (ESU) Algorithm*

The enumerate subgraphs (ESU) algorithm enumerates all size-k subgraphs and discards some of the subgraphs during its execution to have an unbiased subgraph sampling algorithm. ESU starts with a vertex v and adds new vertices to V_{ext} which is the extension set, that have two properties. The label of the new vertex u should be greater than that of v; and it should be adjacent to the newly added vertex v but not to any other vertex in the current subgraph V_s, as shown in Alg. 9.7.

When a new node is selected for expansion, it is removed from the possible extensions and its exclusive neighbors are added to the new possible extensions. This way, the algorithm ensures that each subgraph will be enumerated exactly once since the non-exclusive nodes will be considered in the execution of another recursion. Figure 9.9 displays the execution of this algorithm in a sample graph of six nodes labeled 1,...,6 and the output consists of six triads and a triangle as shown. The ESU algorithm is implemented in the tool FANMOD and uses the *Nauty* algorithm we have seen in Section 9.3 to calculate isomorphisms of graphs [33].

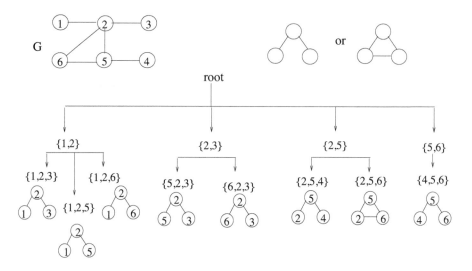

Figure 9.8: Motifs found in a graph using *m finder* **algorithm**

Algorithm 9.7 *ESU*

1: **Input** : $G(V,E)$, **int** $1 \le k \le n$
2: **Output** : All k-size subgraphs of G
3: **for all** $v \in V$ **do**
4:    $V_{ext} \leftarrow \{u \in N(v) : u > v\}$
5:    $ExtSubgraph(\{v\}, V_{ext}, v)$
6: **end for**
7: **return**
8: **procedure** EXTSUBGRAPH(V_s, V_{ext}, v)
9:    **if** $|V_s| = k$ **then output** $G[V_s]$
10:       **return**
11:    **end if**
12:    **while** $V_{ext} \neq \emptyset$ **do**
13:       $V_{ext} \leftarrow V_{ext} \setminus \{\text{an arbitrary vertex } w \in V_{ext}\}$
14:       $V'_{ext} \leftarrow V_{ext} \cup \{u \in N_{excl}(w, V_s) : u > v\}$
15:       $ExtSubgraph((V_s \cup \{w\}, V'_{ext}, v))$
16:    **end while**
17:    **return**
18: **end procedure**

9.4.1.3 Grochow and Kellis Algorithm

Grochow and Kellis proposed an algorithm which searches for a single query graph in the target graph instead of searching all motifs of a given size. In this sense, their

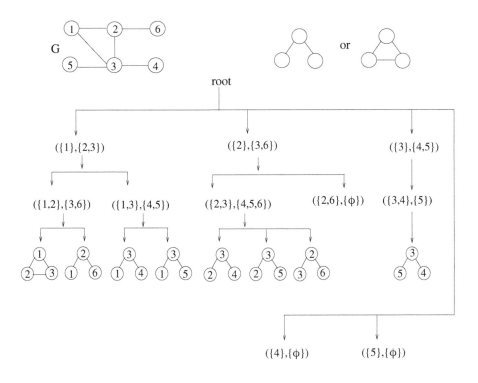

Figure 9.9: Motifs found in a graph using *ESU* algorithm

algorithm is closely related to the subgraph isomorphism problem we have seen in Section 9.3. They used mapping instead of enumerating where the query graph H is mapped to target graph G in every possible way. They also developed a special method to take advantage of subgraph symmetries. Other than these motif centric methods, they considered degrees of nodes and the degrees of their neighbors to improve isomorphism tests. Also, using *subgraph hashing* where subgraphs of a given size are hashed based on their degree sequences provided further significant improvements in the isomorphism tests. Alg. 9.8 shows the pseudocode for this algorithm adapted from [7], where all of the instances of the query graph H is found in the target graph G.

IsoExt procedure shown in Alg. 9.9 uses the backtracking method that finds all of the isomorphisms from H to G. It first selects the vertices with the highest number of already mapped neighbors and then finds the vertices with the highest degrees and the greatest degree sequences among these vertices. Each call to this procedure results in the partial mapping f being enlarged by a single vertex and whenever an extension is made, the newly discovered and mapped node is connected to the already

---

Algorithm 9.8 *Grochow_Alg*

1: **procedure** FINDSUBINST($H(V_1, E_1), G(V_2, E_2)$)
2:    **Input** : H : query graph, G : target graph
3:    **Output** : all instances of H in G
4:    **find** Auth(H)   ▷ find automorphisms of H
5:    $H_E \leftarrow$ equivalence representatives of H
6:    $C \leftarrow$ symmetry breaking conditions for H using H_E and Auth(H)
7:    **order** nodes of G by first by increasing degrees and then increasing neighbor degrees
8:    **for all** $u \in V_2$ **do**
9:       **for all** $v \in H[H_E]$ such that u can support v **do**
10:          $f \leftarrow$ partial map where $f(v) = u$
11:          $IsoExt(f, H, G[C(h)])$ ▷ find all isomorphic extensions of f until symmetry
12:          **add** the images of these maps to instances
13:       **end for**
14:       $G \leftarrow G - \{u\}$
15:    **end for**
16:    **return** the set of all instances
17: **end procedure**

---

mapped nodes. This process guarantees that the returned map is an isomorphism. In order to prevent multiple mappings of the query graph H to a distinct subgraph G' of G, several symmetry breaking conditions are tested to provide a unique mapping from H to G'.

9.4.1.4   Kavosh Algorithm

Kavosh is a recent algorithm proposed by Kashani et al. to detect motifs in undirected and directed graphs [8]. It consists of four subtasks as *enumeration* where all subgraphs of a given size are found; *classification* where each subgraph detected is classified into isomorphic groups; *random graph generation* in which similar random networks to the input network are generated, and *motif identification* where motifs are identified based on statistical evaluations. Enumeration and classification steps are also performed on the random networks.

In the enumeration step, a tree structure with restrictions is formed such that each subgraph is enumerated only once. All subgraphs that include a vertex v are found and vertex v is then removed from the graph, and the procedure is repeated for all vertices in this step. In order to find all subgraphs that contain v, trees rooted at v and having a maximum depth of k are constructed. The first rule to form children of a vertex is that they should have not have been covered in higher levels of the tree. The second rule imposes that all children in a particular tree must have identifiers greater than that of the root. Using these rules, a tree rooted at vertex v is formed by going to the lowest possible level. The tree is then ascended and the process is repeated where

---

Algorithm 9.9 *Iso_Ext*

1: **procedure** ISOEXT($f, H(V_1, E_1), G(V_2, E_2)$)
2:     $D \leftarrow$ domain of f
3:     **if** $D = H$ **then**
4:         **return** f
5:     **else**
6:         $m \leftarrow$ most constrained neighbor of any $u \in D$
7:         **for all** neighbor v of $f(D)$ **do**
8:             **if** $\exists u \in D$ neighbor of $m : v \notin N(f(D))$; or $\exists u \in D$ non-neighbor of $m : v \in N(f(D))$; or $f(m) = u$ violates a symmetry breaking condition in $C(h)$ **then**
9:                 continue with next u
10:            **else**
11:                $f' \leftarrow f$ on D
12:                $f'(m) \leftarrow v$
13:            **end if**
14:            **find** all isomorphic extensions to f'
15:            **append** these maps to isomorphism list
16:        **end for**
17:        **return** the list of isomorphisms
18:    **end if**
19: **end procedure**

---

vertices visited in the previous paths of descendent are now considered unvisited vertices.

In order to extract subgraphs of size k using the tree structure, all possible positive integer sums yielding the $k - 1$ value are considered. For example, to inspect size 4 motifs, Kavosh considers (1,1,1); (1,2); (2,1); and (3) combinations which all give $k - 1$ value as summation. The pseudocode of the enumeration algorithm of *Kavosh* is shown in Alg. 9.10 as adapted from [8]. A vertex v is labeled as *visited* if it is a neighbor to any of the elected vertices in the upper levels of the tree. The set $S_i(i = 0, ..., m), m \leq k - 1$ includes all vertices from the ith level in a subgraph. The *EnumVertex* recursive procedure enumerates all vertices from a specific root vertex and the *Validate* function provides a list of valid vertices to select. The *InitialComb* and *NextComb* functions use the revolving door ordering method [11] to generate a combination of vertices.

The implementation of *Kavosh* enumeration algorithm is shown in Figure 9.10 where $k = 4$ and vertex 1 is the root. The starting composition is (1,1,1) and the vertices 2, 3, and 5 are the valid children of vertex 1 which are marked as visited. Using the revolving door ordering algorithm, first vertex 2 and then vertices 6 and 4 are selected which make up the subgraph consisting of vertices 1, 2, 6 and 4 as shown by bold circles in a.1 in the figure. The tree is ascended recursively without visiting the lower vertices anymore. The composition (1,2) detects the subgraph of

---

Algorithm 9.10 *Kavosh*

1: **Input** : $G(V, E)$ undirected or directed
2: **Output** : L : List of all k-size subgraphs of G
3: **for all** $v \in V$ **do**
4:     $visited[v] \leftarrow true$; $S_0 \leftarrow v$
5:     $EnumVertex(G, v, S, k-1, 1)$
6:     $visited[v] \leftarrow true$
7: **end for**
8: **procedure** ENUMVERTEX(G, v, S, rem)
9:     **if** $rem = 0$ **then**
10:         **return**
11:     **else**
12:         $List \leftarrow Validate(G, S_{i-1}, v)$
13:         $n_i \leftarrow min(|List|, rem)$
14:         **for** $k_i = 1$ to n_i **do**
15:             $C \leftarrow InitialComb(List, k_i)$
16:             **repeat**
17:                 $S_i \leftarrow C$
18:                 $EnumVertex(G, v, S, rem - k_i, i + 1)$
19:                 $NextComb(List, k_i)$
20:             **until** $C = \emptyset$
21:         **end for**
22:         **for all** $u \in List$ **do**
23:             $visited[u] \leftarrow false$
24:         **end for**
25:     **end if**
26: **end procedure**

---

vertices $\{1,2,6,7\}$ (b) first and the composition $(2,1)$ finds $\{1,2,3,6\}$ (c) and finally the last composition (3) detects the subgraph $\{1,2,3,5\}$ in d.

Kavosh uses the *Nauty* algorithm [14] we have seen in Section 9.3, for the classification phase. The random graphs are generated by preserving the degree sequence of the original graph and by switching the edges of the randomly selected vertices. In order to determine whether a found subgraph is a motif or not, Kavosh evaluates the frequency of the motif using the *Z-score* and the *P-value* parameters. Using these procedures, it has successfully detected motifs of sizes 4, 5 and 9 in both biological networks such as the metabolic pathway of the bacteria *E. coli* and the transcription network of yeast, and non-biological entities such as a social network and an electronic network.

9.4.1.5 MODA

MODA algorithm proposed by Omidi et al. [18] uses a pattern growth approach by taking an input query graph and finding the frequency of the occurrences of

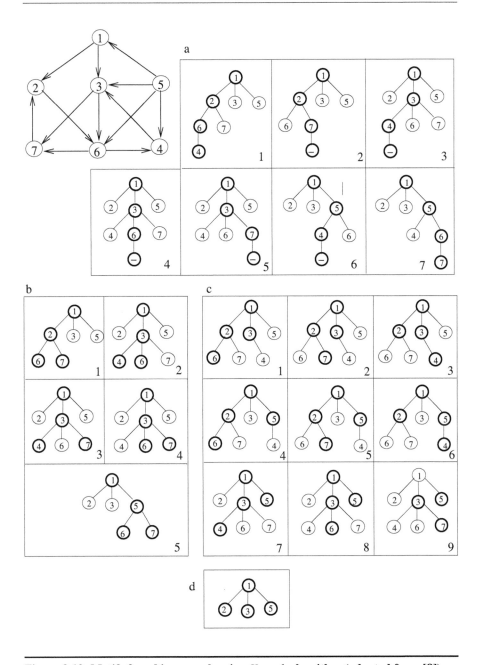

Figure 9.10: Motifs found in a graph using *Kavosh* algorithm (adapted from [8])

the query graph in the target network. If the input query graph is a supergraph of a previous query graph, the previous mapping information is used to reduce computation time. The algorithm uses the concept of expansion trees T_k to prepare a hierarchical structure in the algorithm. The expansion tree T_k has the following properties [18]

1. Level of the root is zero and each node at each level includes a graph of size k with $k - 2 + 1$ edges.

2. Number of nodes at the first level is equal to the number of non-isomorphic trees of size k.

3. Each graph at a node is non-isomorphic to all other graphs in T_k.

4. Each node except the root is a subgraph of its child.

5. The unique leaf node at level $(k^2 - 3k + 4)/2$ holds the complete graph K_k.

6. Longest path from the root node to the leaf has $(k^2 - 3k + 4)/2$ edges.

It starts by calculating the frequency of trees of the required size k in the network and expanding them by adding edges until a complete graph of size k is obtained. Each node in the expansion tree has the adjacency matrix of the graph it represents. In order to generate the expansion tree consisting of a set of non-isomorphic trees of size k, distinct trees at the first level are identified. The trees and later graphs are then expanded by the manipulation of the adjacency matrices and this process continues until K_k is obtained. In the second step, the DFS algorithm is executed in this directed graph to obtain the expansion tree. The expansion trees are formed for all graph sizes once and can be used during algorithm execution. Finally, the subgraph frequencies are calculated using *mapping* and *enumeration* modules. The pseudocode of the MODA subgraph frequency finding algorithm is shown in Alg. 9.11 [18, 24].

Get_Next_BFS function performs a BFS traversal the expansion tree T_k and the two functions *MappingModule* and the *EnumeratingModule* enumerate the occurrences of a query graph in the network. If the selected node by the BFS function in the expansion tree is a tree, *MappingModule* is called which implements the Grochow algorithm, otherwise the *EnumeratingModule* function is called which loads the graphs for the parent nodes from memory and tests whether the new edge of the expansion tree exists in the original network for all possible graphs.

The expansion tree T_4 for 4-node graphs is shown in Figure 9.11. An edge is added at each level to form a tree at the next (child) level, therefore, all of the graphs at each level are non-isomorphic, preventing redundancy. The depth of the tree T_k is determined by a node, when a complete graph of k nodes (K_k) is formed at that node. MODA requires substantial memory space as all k-subgraphs are stored in memory.

9.4.2 Approximate Algorithms with Sampling

Since the number of subgraphs grows exponentially both with the size of the network and the size of the subgraph investigated, the exact census algorithms demand high computational power and their performance degrades with the increased motif size. An efficient way to overcome this problem is to employ probabilistic approximation

---

Algorithm 9.11 *MODA*

1: **Input** : $G(V,E)$, **int** $1 \leq k \leq n$, Δ : threshold value
2: **Output** : L : List of frequent k-size subgraphs of G
3: **repeat**
4:     $G'(V',E') \leftarrow Get\_Next\_BFS(T_k)$ ▷
5:     **if** $|E'| = k - 1$ **then**
6:         $MappingModule(G',G)$
7:     **else**
8:         $EnumeratingModule(G',G,T_k)$
9:     **end if**
10:    save F_2
11:    **if** $|F_G| > \Delta$ **then**
12:        $L \leftarrow L \cup G'$ ▷ add subgraph to the list
13:    **end if**
14: **until** $|E'| = (k-1)/2$
15: **procedure** ENUMERATINGMODULE(G',G,T_k)
16:     $F_G \leftarrow \emptyset$
17:     $H \leftarrow Get\_Parent(G',T_k)$
18:     **get** F_H from memory
19:     $(u,v) \leftarrow E(G') - E(H)$
20:     **for all** $f \in F_H$ **do**
21:         **if** $f(u),f(v) \in G$ **and** $< f(u),f(v) >$ violates the corresponding conditions **then**
22:             **add** f to F'_G
23:         **end if**
24:     **end for**
25:     **return** F_G
26: **end procedure**
27: **procedure** MAPPINGMODULE(G,G')
28:     $Grochow\_Kellis(G',G)$
29: **end procedure**

---

algorithms where a number of subgraphs of required size are sampled in the target graph and in randomly generated graphs, and the algorithms are executed in these samples rather than the entire graphs. The accuracy of the method employed will increase with the number of samples used, however, we need to ensure that these subgraphs are qualitative representatives of the original graphs to provide the results with reasonable certainty. We will describe three such algorithms in this section which are the random versions of the *mfinder*, ESU and MODA algorithms.

9.4.2.1 *M finder with Sampling*

The randomized version of *mfinder* starts by selecting a random edge (u,v) and includes the vertices in this edge in the subgraph vertex set V_s [10]. It then selects

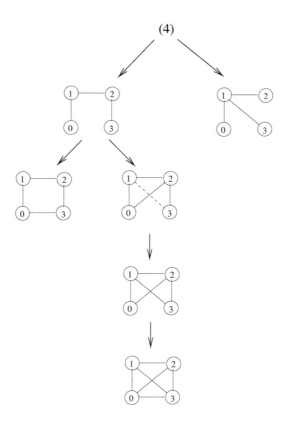

Figure 9.11: Motif search using *MODA* algorithm (adapted from [18])

edges around this edge randomly and adds vertices that are neighbors to vertices in V_s but not already included in V_s to V_s. This process continues until the required motif size k is reached as shown in Alg. 9.12. In the end, the probability to sample V_s is also returned. This algorithm may be called as many times as required to obtain approximate census.

9.4.2.2 Randomized ESU Algorithm

The probabilistic approximate version of the algorithm ESU is similar in structure to the original ESU algorithm [31]. Since complete traversal is time costly, only parts of the ESU tree can be searched such that each leaf is accessed with equal probability. To achieve this, a probability P_d is assigned for each depth $1 \leq d \leq k$ in the tree and the subtree rooted at each child node in the tree is traversed with this probability. This is shown by the addition of "With probability P_d" in line 5 and "With probability $P_{|V'_s|}$" in line 17 of the original algorithm as shown in Alg. 9.13. The probability of finding subgraphs near the searched vertex is higher than finding remote ones, therefore, P_d should be inversely proportional to depth of the enumeration d.

Algorithm 9.12 *EdgeSampling_Alg*

1: **procedure** EDGESAMPLE(G, k)
2:    **Input** : $G(V, E)$,   ▷ weighted or unweighted
3:         **int** $2 \le k \le n$
4:    **Output** : A subgraph of size k
5:    $E_s \leftarrow \emptyset; V_s \leftarrow \emptyset$
6:    **pick** a random edge $(u, v) \in E$
7:    $E_s \leftarrow (u, v); V_s \leftarrow \{u, v\}$
8:    **while** $|V_s| \ne k$ **do**
9:       $L \leftarrow$ neighbor edges of $\{u, v\}$
10:       $L \leftarrow L \backslash \{$ all edges between the vertices in $V_s\}$
11:       **if** $L = \emptyset$ **then exit**
12:       **end if**
13:       **pick** a random edge $(w, z) \in L$
14:       $V_s \leftarrow V_s \cup \{w, z\}$
15:       $E_s \leftarrow E_s \cup (w, z)$
16:    **end while**
17:    **return** V_s
18: **end procedure**

Algorithm 9.13 *ESU_Alg*

1: **Input** : $G(V, E)$, **int** $1 \le k \le n$
2: **Output** : All k-size subgraphs of G
3: **for all** $v \in V$ **do**
4:    $V_{ext} \leftarrow \{u \in N(v) : u > v\}$
5:    With probability P_d
6:       $Ext\_Subgraph(\{v\}, V_{ext}, v)$
7: **end for**
8: **return**
9: **procedure** $Ext\_Subgraph(V_s, V_{ext}, v)$
10:    **if** $|V_s| = k$ **then output** $G[V_s]$
11:       **return**
12:    **end if**
13:    **while** $V_{ext} \ne \emptyset$ **do**
14:       $V_{ext} \leftarrow V_{ext} \backslash \{$an arbitrary vertex $w \in V_{ext}\}$
15:       $V_{ext} \leftarrow V_{ext} \cup \{u \in N_{excl}(w, V_s) : u > v\}$
16:       $V'_s \leftarrow V_s \cup \{w\}$
17:       With probability $P_{|V'_s|}$
18:          $Ext\_Subgraph((V_s \cup \{w\}, V'_{ext}, v))$
19:    **end while**
20:    **return**
21: **end procedure**

9.4.2.3 MODA with Sampling

The mapping module of MODA uses a lot of time although it searches a small part of the expansion tree. A sampling method is added to this module to improve the performance which searches only the samples based on the root vertices to reduce time [18]. The new mapping module uses a probability distribution proportional to the degrees of the nodes in line 5, since probability of finding a subgraph around higher degree nodes is higher than the ones with lower degrees as shown in Alg. 9.14 [24].

Algorithm 9.14 *MODA_Sample*

1: **Input** : network graph $G(V,E)$, query graph $G'(V',E')$
2: **Output** : approximate frequency and mapping of G'
3: **procedure** MAPSAMPLE(G')
4:     **for** $i = 1$ to $n_{s}amples$ **do**
5:         **select** $u \in V$ with probability $deg(v)$
6:         **for all** $v \in V'$ **do**
7:             **if** $deg(u) \geq deg(v)$ **then**
8:                 $Grochow(G')$ with $f(v) = u$
9:             **end if**
10:            $V \leftarrow V \setminus \{u\}$
11:        **end for**
12:    **end for**
13: **end procedure**

9.5 Chapter Notes

In this chapter, we first defined the network motif discovery problem and showed it is closely related to the subgraph isomorphism problem. We then described three algorithms due to Ullman [29], McCay [15] and Cordelle et al. [4] that are used to find isomorphic subgraphs of a graph. The algorithm due to Ullman and the *Nauty* algorithm is still used in various motif detecting algorithms such as *mfinder* and Kavosh. Graph isomorphism is a well-studied topic in mathematics and computer science and the fastest known algorithm has $2^{\sqrt{n \log n}}$ complexity for general graphs [12]. For special graph classes such as bounded degree graphs, polynomial time algorithms can be found [13]. For graph or subgraph isomorphism, approximate algorithms using heuristics provide polynomial time solutions and may be preferred when motifs of large sizes, greater than 5 or 6, are searched.

The first algorithm to find network motifs called *mfinder* is due to Milo et al. [9] which uses exhaustive enumeration of all subgraphs. It may discover the same motif several times and due to its excessive usage of memory space, its implementation is limited to motifs of size 5 or less. Kashtan et al. proposed a procedure based on *mfinder* to find motifs using a probabilistic approach which can find motifs in reasonable time which we called *sampled mfinder* [9]. Wernicke introduced a fast algorithm that is used in the tool FANMOD which can discover motifs up to size of 8 efficiently using unbiased sampling where each motif is discovered exactly once. Grochow et al. also proposed an algorithm that uses an algebraic method to discover motifs [7], which employs biased sampling.

Kavosh algorithm proposed by Kashani et al. finds all occurrences of subgraphs that include a specific node [8]. Omidi et al. introduced the MODA algorithm which uses a pattern growth approach by starting from a single node and extending it until a complete graph with *k*-nodes is formed [18]. Riberio et al. compared the efficiency of *mfinder*, FANMOD and Grochow algorithms in circuit, yeast and social networks [23]. They concluded that *mfinder* is slower than FANMOD on all networks and all sizes and also its requirement of large memory space prevents its usage for motifs of sizes larger than 5. In their experiments, Grochow algorithm performed better than FANMOD for undirected graphs. Also, if the motif to be searched is known beforehand, Grochow outperformed FANMOD in large motif detection.

There are only few research studies in the study of parallel algorithms for motif detection. Wang et al. used static load balancing based on node degrees prior to computation to distribute load evenly [30]. Schatz et al. attempted to parallelize the Grochow-Kellis algorithm by performing single graph queries on a number of processors [26]. Ribeiro proposed a novel data structure called *g-tries* and a sampling method based on *g-tries* in his Ph.D. thesis [24] where parallel algorithms based on *g-tries* and ESU are provided. We see that there are only few parallel algorithms for this computationally heavy task and therefore can conclude it is a potential area that should be researched more.

Exercises

1. Find the motif shown in Figure 9.12 in the graph shown in (b) of the same figure using F_1, F_2 and F_3 frequency concepts.

2. Show that the two graphs shown in Figure 9.13 are isomorphic by finding a permutation matrix P that transforms one to the other.

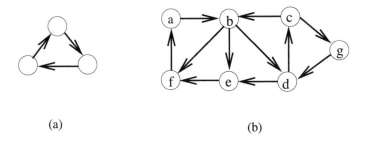

(a)   (b)

Figure 9.12: Example graph for Ex. 1

Figure 9.13: Example graph for Ex. 2

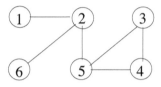

Figure 9.14: Example graph for Ex. 3

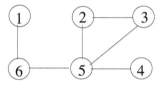

Figure 9.15: Example graph for Ex. 4

3. Find the 3-node network motifs in the graph of Figure 9.14 using the *mfinder* algorithm.

4. Find the 3-node network motifs in the graph of Figure 9.15 using the *ESU* algorithm.

5. Find the 3-node network motifs in the graph of Figure 9.16 using the Kavosh algorithm.

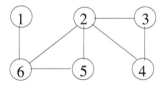

Figure 9.16: Example graph for Ex. 5

6. Compare exact census algorithms with approximate census algorithms in terms of their performances and the precision of the results obtained in both cases.

References

[1] I. Albert and R. Albert, Conserved network motifs allow protein-protein interaction prediction. *Bioinformatics*, 20(18):3346-3352, 2004.

[2] R. Battiti and F. Mascia. An algorithm portfolio for the subgraph isomorphism problem. In *Engineering Stochastic Local Search Algorithms. Designing, Implementing and Analyzing Effective Heuristics*, pages 106-120. Springer, Berlin, 2007.

[3] L.P. Cordella, P. Foggia, C. Sansone, and M. Vento. Evaluating performance of the VF graph matching algorithm. In *Proceedings of the 10th Intl Conf. Image Analysis and Processing*, pages 1172-1177, 1999.

[4] L.P. Cordella, P. Foggia, C. Sansone, M. Vento. (2004). A (sub) graph isomorphism algorithm for matching large graphs. *IEEE Transactions on Pattern Analysis and Machine Intelligence*, 26(10):1367-1372, 2004.

[5] S. Fortin. The Graph Isomorphism Problem. Technical Report TR 96-20, July 1996, Dept. of Computer Science, The University of Alberta.

[6] M.R. Garey and D.S. Johnson. *Computers and Intractability: A Guide to the Theory of NP-Completeness*. W. H. Freeman, 1979.

[7] J. Grochow and M. Kellis. Network motif discovery using subgraph enumeration and symmetry-breaking. In *Proceedings of the 11th annual international Conference on Research in Computational Molecular Biology, RECOMB'07*, pages 92-106, 2007.

[8] Z.R. Kashani, H. Ahrabian, E. Elahi, A. Nowzari-Dalini, E.S. Ansari, S. Asadi, S. Mohammadi, F. Schreiber, A. Masoudi-Nejad. Kavosh: a new algorithm for finding network motifs. *BMC Bioinformatics*, 10(318), 2009.

[9] N. Kashtan, S. Itzkovitz, R. Milo, and U. Alon. Mfinder tool guide. Technical Report, Department of Molecular Cell Biology and Computer Science and Applied Mathematics, Weizmann Institute of Science, 2002.

[10] N. Kashtan, S. Itzkovitz, R. Milo, and U. Alon. Efficient sampling algorithm for estimating subgraph concentrations and detecting network motifs. *Bioinformatics*, 20:1746-1758, 2004.

[11] D. Kreher, D. Stinson. *Combinatorial Algorithms: Generation, Enumeration and Search*, CRC Press, 1998.

[12] L. Babai and E. M. Luks. Canonical labeling of graphs. In *Proceedings of the Fifteenth Annual ACM Symposium on Theory of Computing, STOC'83*, pages 171-183, 1983.

[13] Eugene M. Luks. Isomorphism of graphs of bounded valence can be tested in polynomial time. *J. Comput. Syst. Sci.*, 25(1):42-65, 1982.

[14] Nauty User's Guide, Computer Science Dept. Australian National University.

[15] B. D. McKay, Practical graph isomorphism. In *Manitoba Conference on Numerical Mathematics and Computing Winnipeg 1980*, Congressus Numerantium, 30:45-87, 1981.

[16] B.T. Messmer, H. Bunke. Subgraph isomorphism detection in polynomial time on preprocessed model graphs. In *Recent Developments in Computer Vision*, pages 373-382, Springer, Berlin, 1996.

[17] H. Schwbbermeyer. Network Motifs. Junker BH, Schreiber F (Eds). *Analysis of Biological Networks*. John Wiley and Sons., pages 85-108, 2008.

[18] S. Omidi, F. Schreiber, A. Masoudi-Nejad. MODA: an efficient algorithm for network motif discovery in biological networks. *Genes and Genetic Systems*, 84:385-395, 2009.

[19] N. Kashtan, S. Itzkovitz, R. Milo, and U. Alon. Mfinder tool guide. Technical Report, Department of Molecular Cell Biology and Computer Science and Applied Mathematics, Weizmann Institute of Science, 2002.

[20] R. Milo, S. Shen-Orr, S. Itzkovitz, N. Kashtan, D. Chklovskii, and U. Alon. Network motifs: simple building blocks of complex networks. *Science*, 298(5594):824-827, 2002.

[21] Mfinder. http://www.weizmann.ac.il/mcb/UriAlon/index.html

[22] S. S. Shen-Orr, R. Milo, S. Mangan, and U. Alon, Network motifs in the transcriptional regulation network of escherichia coli, *Nature Genetics*, 31(1):64-68, 2002.

[23] P.M.P. Ribeiro, F. Silva and M. Kaiser. Strategies for Network Motifs Discovery. In *Proceedings of the 5th IEEE International Conference on e-Science (ESCIENCE)*, pages 80-87, IEEE CS Press, Oxford, UK, December, 2009.

[24] P.M.P. Ribeiro, Efficient and Scalable Algorithms for Network Motifs Discovery, PhD Thesis. Doctoral Programme in Computer Science. Faculty of Science of the University of Porto, 2011.

[25] Esra Ruzgar, A Survey on Exact Algorithms for Subgraph Isomorphism Problem, unpublished report.

[26] M. Schatz, E. Cooper-Balis, and A. Bazinet. Parallel network motif finding, CMSC 714 High Performance Computing, 2008.

[27] F. Schreiber and H. Schwbbermeyer. Frequency concepts and pattern detection for the analysis of motifs in networks. *Transactions on Computational Systems Biology III*, LNBI 3737, pages 89-104. Springer-Verlag, Berlin, 2005.

[28] H. Schwbbermeyer. Network Motifs. In: Junker BH, Schreiber F (Eds.). *Analysis of Biological Networks*. Hoboken, New Jersey: John Wiley & Sons, pages 85-108, 2008.

[29] J.R. Ullman. An algorithm for subgraph isomorphism. *Journal of the ACM*, 23(1):31-42, 1976.

[30] Tie Wang, Jeffrey W. Touchman, Weiyi Zhang, Edward B. Suh, and Guoliang Xue. A parallel algorithm for extracting transcription regulatory network motifs. In *Proceedings of the IEEE International Symposium on Bioinformatics and Bioengineering*, pages 193-200, Los Alamitos, CA, USA, 2005. IEEE Computer Society Press.

[31] S. Wernicke. Efficient Detection of Network Motifs. *IEEE/ACM Trans. Comput. Biology Bioinform.* 3(4):347-359, 2006.

[32] S. Wernicke. A Faster Algorithm for Detecting Network Motifs. In *Proceedings of the 5th WABI-05*, Volume 3692, pages 165-177, Springer, 2005.

[33] S. Wernicke and F. Rasche. FANMOD: A tool for fast network motif detection. *Bioinformatics*, 22(9):1152-1153.

APPLICATIONS

Chapter 10

Protein Interaction Networks

10.1 Introduction

Proteins have important functions in cell biology such as transferring signals, controlling the functions of enzymes and regulating activities in the cell. Proteins interact with other proteins in the cell to form protein-protein interaction (PPI) networks to perform important tasks in the cell such as cell cycle control, protein folding, translation and transcription. A protein may also modify another protein by interacting with it in a PPI network. Understanding the role of interactions of proteins is believed to provide molecular indications of health and disease states which can be used to provide new drugs and therapies.

PPI networks can be modeled by graphs where nodes represent proteins and edges show the interactions between them. These networks act as an interface between the genome and the metabolism as shown in Figure 10.1. They perform the functions under the control of the genes which results in biochemical reactions governing metabolism. Using technologies such as mass spectrometry provided large volumes of data of PPI networks; however, the size of data and the fact that it contains significant noise make the analysis difficult.

There are many public PPI databases such as Munich Information center for Protein Sequences (MIPS) [26], Yeast Proteomics Database (YPD) [6], Database of Interacting Proteins (DIP) [39] and Human Reference Protein Database (HRPD) [12] which contain data about PPI networks of various organisms. Recent studies have shown that disease genes share common topological properties [28, 14]. Assessment of the relationship between PPI network topology and biological function of PPI networks and disease is a challenging and active research area.

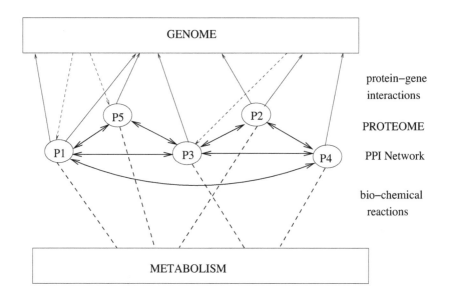

Figure 10.1: Networks in the cell

A part of a protein that has a specific functionality such as signal transduction and transcription is called a *domain* [30]. Functions performed by domains within a protein include signal transduction, transcription and metabolism. The same domain may be found in various proteins. A *molecular pathway* is a sequence of directed molecular reactions to perform a process in the cell. A protein in a cell may regulate the abundance of another, forming a pair with it. A *genetic regulatory network* in a cell is formed by all such pairs of proteins.

In this chapter, we will first review the properties of PPI networks from the graph theory point of view. We will then describe algorithms that are used to find clusters in PPI networks. We will also describe network motifs and network alignment algorithms in these and other biological networks.

10.2   Topological Properties of PPI Networks

Topology of the PPI networks is important to discover functions of unknown proteins, to understand the evolution of protein interaction better and to understand the high level functional organization in the cell. A fundamental aim in studying the topology of the PPI networks therefore is to discover and predict the network structure that is related to the biological functions and processes in the cell. There are other types of interactions in the cell such as the metabolic, signalling and transcription-regulatory networks and all of these networks cooperate to result in the overall

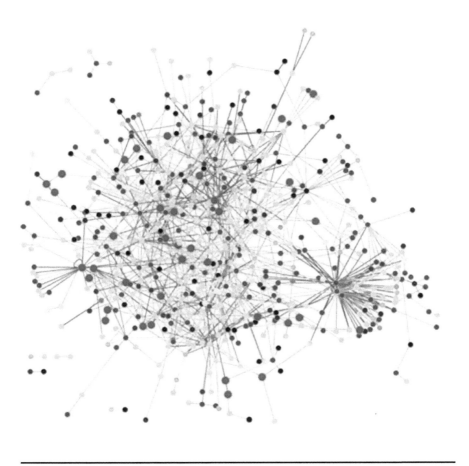

Figure 10.2: The PPI network of *T. pallidum* **taken from [38]**

function of the cell. An example protein interaction network of *Treponema pallidum* including 576 proteins and 991 interactions is shown in Figure 10.2 [38]. Proteins involved in DNA metabolism are shown as enlarged circles.

The structures of the PPI networks have been observed to have a scale-free topology where $P(k) \approx k^{-2.2}$ [30], with a few high degree nodes called *hubs* and many small degree nodes. These hubs are believed to have important roles in the overall functioning of the network. Proteins in a PPI network interact with other proteins in the same functional class but also with other proteins in different functional classes. This property can be best described by the hierarchical clustering we have described in Chapter 7. The clustering coefficient of a vertex u showed how well connected the neighbors of u are and the average clustering coefficient was the average value of the clustering coefficients of all vertices. Studies have shown that the average clustering coefficient in PPI networks is almost constant, that is, it does not change significantly with the size of the network. The average shortest path length is low in PPI networks as found in small world networks [13].

10.3 Detection of Protein Complexes

A group of proteins usually cooperate to perform tasks related to the cell of an organism. These groups which have high connectivity between them are called *protein complexes* and include some core proteins that are always in the complex and other proteins which are in the complex only when needed. The principles of cell organization and function can be better understood if these complexes are discovered. PPI network analysis involves detecting these complexes which in fact is clustering of the PPI networks conceptually.

Although there have been several studies to detect protein complexes using mass spectrometry, computational approaches based on the graph model of the PPI networks are relatively more recent and less investigated. In this section, we will describe fundamental algorithms to detect protein complexes in PPI networks. Although the graph clustering algorithms may also be used for this purpose, the algorithms we will discuss have been designed for biological networks and therefore consider the scale-free structure of these networks and also their performances have been evaluated in these networks in various studies.

10.3.1 Highly Connected Subgraphs Algorithm

The clique-based clustering algorithms we saw in Section 8.4 will only detect cliques and in many graphs, there may be highly connected subgraphs that do not show a clique structure. The highly connected subgraphs (HCS) algorithm due to Hartuv and Shamir [11] consider the connectivity of subgraphs of a graph while finding clusters. We will briefly review the concept of connectivity in graphs before describing this algorithm. The *edge connectivity* $\kappa_E(G)$ of a graph $G(V,E)$ is the minimum number of edges whose removal disconnects G. A *cut* in a graph G is a set of edges whose removal disconnects G, and a minimum cut is a cut with minimum size. Hartuv and Shamir defined a graph as highly connected if $\kappa_E(G) \geq \frac{n}{2}$ and a highly connected subgraph of G as a highly connected induced subgraph of G. The HCS algorithm uses the procedure $MinCut(G)$ which returns H, $\bar{H}$ and C where C is the minimum cut that separates G into subgraphs H and $\bar{H}$. This recursive algorithm either returns a highly connected subgraph or it continues with the recursion as shown in Alg. 10.1 [11].

Single nodes are returned in the singleton set S and the final output consists of clusters returned during the execution of the HCS algorithm. The time complexity of the algorithm has an upper bound of $2N \times f(n,m)$ where N is the number of clusters discovered by the algorithm and $f(n,m)$ is the time complexity of finding a minimum cut in a graph that has n vertices and m edges. The fastest minimum edge cut algorithm currently has $O(nm)$ time complexity [24] for unweighted graphs and

---

Algorithm 10.1 *HCS_Alg*

1: **procedure** *HCS*(*G*)
2:     **Input** : $G(V,E)$
3:     **Output** : highly connected clusters of G
4:     $(H,\bar{H},C) \leftarrow MinCut(G)$
5:     **if** G is highly connected **then**
6:         **return**(G)
7:     **else**
8:         HCS(H)
9:         HCS($\bar{H}$)
10:    **end if**
11: **end procedure**

$O(nm + n^2 \log n)$ for weighted graphs [37]. Figure 10.3 displays the execution of the HCS algorithm in a sample graph which produces three clusters C_1, C_2 and C_3 that all have a connectivity greater than $n/2$.

10.3.2   *Restricted Neighborhood Search Algorithm*

King et. al. developed the restricted neighborhood search (RNSC) algorithm to partition PPI networks into clusters using a cost function which is applied to each partitioning [19]. They applied RNSC successfully to four types of PPI networks in two steps. The RNSC algorithm was used to cluster the network in the first step and the

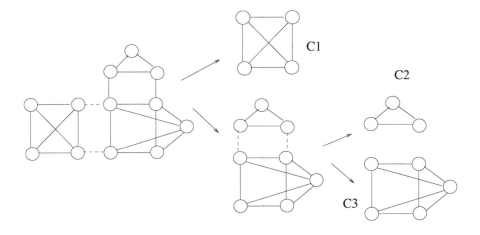

Figure 10.3: Implementation of HCS algorithm

results were filtered based on clustering size, density and functional homogeneity to emphasize properties observed in biological complexes.

In essence, the RNSC algorithm is a cost-based local search algorithm based on the tabu search heuristic [10]. Starting from a random or user-input clustering, it first performs a search using a simple integer valued function called the *naive cost function* as a preprocessor. It then searches the graph using a real valued cost function called the *scaled cost function*. The algorithm operates similar to the Kernighan-Lin algorithm [18] by iteratively moving one node from one cluster to another randomly to improve the cost of the clustering as shown below:

1. Start with a random partitioning.

2. Move a node u from a cluster C_x to another cluster C_y to minimize the naive cost function defined as follows:

   ■ The number of neighbors of u that are not in C_x + the number of nodes in C_x that are not its neighbors.

3. Occasionally, some clusters are destroyed by dispersing their nodes to other clusters at random.

4. If there is no improvement after a certain number of steps, algorithm starts from step 2 using the scaled cost function which is the naive cost function scaled by the size of the network.

It also makes diversification moves by occasionally dispersing the contents of a cluster randomly. The RNSC output is filtered by setting a maximum P-value for functional homogeneity, a minimum density and a minimum size and only the clusters that meet this criteria are output as protein complexes.

10.3.3   Molecular Complex Detection Algorithm

The molecular complex detection (MCODE) algorithm is a graph-theoretic clustering algorithm to identify protein complexes in large PPI networks [3]. This algorithm operates in three stages as vertex weighting, prediction of complexes and optionally post-processing to filter or add proteins to the resulting complexes.

The clustering coefficient of a vertex i was defined as $cc(i) = 2m_i/n_i(n_i - 1)$ where n_i is the number of neighbors of i and m_i is the number of edges between these neighbors. This parameter is the ratio of the existing connection between the neighbors of a vertex to the maximum possible connections between them, and shows how well connected the neighbors of a vertex are. The density of a graph G is the ratio of the number of edges of G to the number of maximum possible edges G may have, as $\rho = 2m/n(n - 1)$ and is a real number between 0 and 1.0.

A k-core subgraph of a graph G in MCODE is an induced subgraph G' of G where any vertex in G' has at least a degree of k. An example 3-core including vertex u is shown by gray vertices in Figure 10.4. Based on these definitions, the *core-clustering coefficient* of a vertex v is defined as the density of the highest k-core of

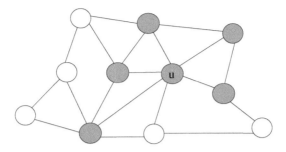

Figure 10.4: k-core example

the closed neighborhood of v. The core-clustering coefficient of u in this figure is 0.6. This parameter is used to emphasize the dense regions of the graph G. Since PPI networks are known to be scale free, they exhibit few high degree nodes with many low degree nodes.

In the first phase of the MCODE algorithm, each vertex u is assigned a weight which is equal to the product of its core-clustering and the highest k-core value in its closed neighborhood. The weight for vertex u in Figure 10.4 found in this manner is $3 \times 0.6 = 1.8$. This way, we are emphasizing the density and the degree of its neighbors. In the second phase, the highest weight vertex v is selected as the seed of the initial cluster and a BFS is performed from this seed. A vertex v is included in the cluster if it is the neighbor of u and $w(v) > TWP$ where TWP is the threshold weight value to be included in the cluster. A cluster that cannot be enlarged any further is removed from the graph and the procedure is repeated until no more clusters are found. In the optional post-processing phase, the following actions can be performed:

- **Filter**: Clusters that do not contain a 2-core are discarded. Also, nodes with very low weights are removed from clusters.

- **Flush**: For every node u in a cluster C, if the density of $\{u\} \cup N(u)$ exceeds a threshold value, the nodes in $N(u)$ are added to C, if they are not part of other clusters to prevent overlaps.

- **Haircut**: The tree-like structures are removed from the clusters.

10.3.4 Markov Clustering Algorithm

The Markov clustering algorithm (MCL) is an effective clustering method designed for PPI networks. It was proposed by van Dongen [8] based on the following heuristics:

1. Number of paths of length k between two nodes u and v in a graph G is larger if u and v are in the same cluster and smaller if they are in different clusters.

2. A *random walk* starting from a vertex in a dense cluster of G will likely end in the same dense cluster of G.

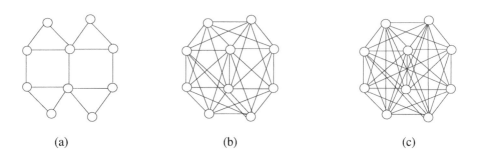

(a)　　　　　　　　　　　(b)　　　　　　　　　　　(c)

Figure 10.5: Powers of a graph G**, a)** G**, b)** G^2**, c)** G^3

3. Edges between clusters are likely to be incident on many shortest paths between all pairs of nodes.

The algorithm is based on the assumption that by doing the random walks on the graph, we may be able to find where the flow gathers which shows where the clusters are. Random walks on the graph are found using the *Markov chains*. A Markov chain has a sequence of variables representing states where the probabilities for the next time step depend only on the current probabilities. Based on these assumptions, the MCL algorithm takes the adjacency matrix A of the graph representing the PPI network as its input. It also normalizes and transforms A into the probability matrix A_P which has entries a_{ij} showing the probability of reaching vertex j from vertex i. It then performs two basic operations called *expansion* and *inflation*. In each step of the main loop of the algorithm, A_P is expanded by taking its eth power and the resulting matrix is inflated by the parameter r.

The kth power of an undirected graph G is the graph G^k which has the same vertex set as G and there is a unity entry in G^k between the vertices u and v if distance between them in G is at most k. Figure 10.5 displays the square and the cube of a graph.

The kth power of a graph G can therefore be defined as the graph G^k with an adjacency matrix A_{G^k} which has the sum of entries of adjacency matrix A_G of G as follows:

$$A_{G^k} = \sum_{i=1}^{k} [A_G]^i \qquad (10.1)$$

which shows the number of all paths of length up to k [33]. The dth power of a graph with a diameter d is a complete graph K_d. In the second step of inflation, the higher connected neighbors of a vertex u are further strengthened and the less connected neighbors of u are weakened by taking powers of specific columns and re-normalizing. The inflation parameter r is used to control the degree of strengthening and weakening. Expansion provides new probabilities for a node pair u and v which are the endpoints of a random walk. Inflation on the other hand, modifies the probabilities for all walks. The algorithm is specified in pseudocode in Alg. 10.2.

Algorithm 10.2 *MCL_Alg*

 1: **Input** : $G(V,E)$ ▷ undirected graph
 2:     expansion parameter e
 3:     inflation parameter r
 4: **Output** : Clusters $C_1,...,C_k$ of G
 5: $A \leftarrow$ adjacency matrix of G
 6: **normalize** A
 7: **form** the Markov Chain probability matrix A_P
 8: $A_{inf} \leftarrow A_P$ ▷ initialize
 9: $k \leftarrow 1$
10: **while** *steady_state* not reached **do**
11:    $A_{exp} \leftarrow A_{inf}^e$ ▷ expand
12:    **inflate** A_{exp} to A_{inf} using r ▷ inflate
13:    $k \leftarrow k+1$
14: **end while**
15: **interpret** the resulting matrix A_{inf} to find clusters

---

MCL has been successfully used for PPI networks in various studies [4, 40]. It requires $O(n^3)$ steps since multiplication of two $n \times n$ matrices during expansion takes n^3 steps, and the inflation can be performed in $O(n^2)$ steps. The convergence of this algorithm has been shown experimentally only where the number of rounds required to converge was between 10 to 100 steps [3].

10.4 Network Motifs in PPI Networks

Network motifs were defined as subgraph structures that appear significantly higher than expected in a random graph. In order to consider a repeating pattern as a motif, the occurrence of this subgraph in a random network should be evaluated and the generation of a number of appropriate random graphs is needed. A general principle is to consider a subgraph pattern as a motif if the probability of this pattern appearing in a random network of the same size is lower than 0.01 [30] as we have investigated in Chapter 9. A fundamental observation is that the occurrence of motifs in real networks grows linearly with the size of the network while it decreases significantly in random networks.

Analysis of the transcriptional regulation network of *Escherichia coli* showed that it contains three basic motifs as follows [30, 35]:

1. Feedforward loop

2. Single input module (SIM)

3. Dense overlapping regions (DOR)

The feedforward loop shown in Figure 10.6.a functions as a circuit that only responds to persistent signals while rejecting transient activation signals [35]. Nodes

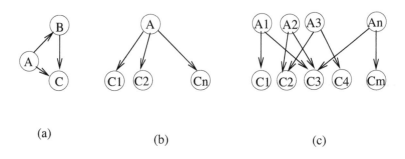

(a) (b) (c)

Figure 10.6: Motifs in *E. coli* **transcriptional regulation network. a) Feedforward loop. b) SIM. c) DOR**

A and *B* in function as the two inputs of an AND gate to control the node *C*. The SIM shown in (b) provides the activation by different genes with different activation thresholds to be temporally ordered. In general, different networks have different motifs.

Other frequently found motifs in biological networks called *bifan*, *biparallel* and *three chain* are shown in Figure 10.7. Bifan is found in both gene regulatory networks and neural networks; biparallel is found in neural networks and the food web; and three chain is found in the food web.

Przujl introduced the term *graphlet* for a small, connected network [30]. The graphlets for 3, 4 and 5 nodes are shown in Figure 10.8. They found the occurrences of all the graphlets in four PPI networks. The relative frequency of the graphlets is defined as $N_i(G)/T(G)$ where $N_i(G)$ is the number of occurrences of graphlet of size i with $i \in \{1,...,29\}$ in graph G; and $T(G) = \sum_{i=1}^{29} N_i(G)$ is the total number of all graphlets in G. The *similarity* between two graphs then should only depend on the differences between the relative frequencies of the graphlets. They defined the *relative graphlet frequency distance* between two graphs G_1 and G_2 as:

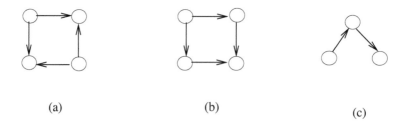

(a) (b) (c)

Figure 10.7: Motifs in biological networks. a) Bifan. b) Biparallel. c) Three-chain motifs

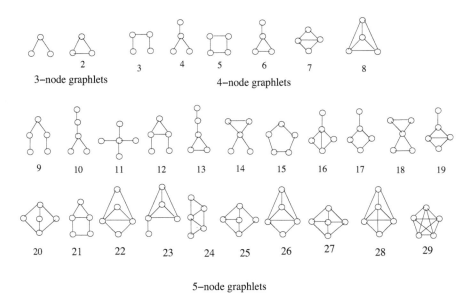

Figure 10.8: Graphlets of size 3, 4 and 5 (adapted from [30])

$$D(G_1, G_2) = \sum_{i=1}^{29} |F_i(G_1) - F_i(G_2)| \qquad (10.2)$$

where $F_i(G) = -\log N_i(G)/T(G)$. Testing the high confidence PPI networks using this parameter, they concluded that the graphlet frequency distributions of these networks were close to the graphlet frequency distributions of geometric random networks. When noise in the PPI networks was considered, the graphlet distribution and the global parameters of these networks were similar to those of scale-free networks. Since searching for all types of graphlets in a large network as a PPI network is time consuming, heuristic algorithms are frequently used as we have seen in Chapter 9.

NeMoFinder proposed by Chen et al. is a network motif discovery algorithm targeting PPI networks [6]. The algorithm inputs a PPI network G, a frequency threshold F, a uniqueness threshold S, and a maximal network motif size k and outputs a set U of repeated motifs of sizes between 3 and k. It consists of three main steps; in the first step, the repeated subgraphs of the PPI network are detected. Then, the frequencies of these repeated subgraphs are determined in randomized networks. Finally, the uniqueness of the subgraphs is tested in the last step.

The first step is divided into three substeps where repeated size-k subtrees as found in the first substep. Trees of size 2 are detected and enlarged to trees of 3, 4 and more, and trees up to size k are found. When a size-k tree is found, its frequency is also calculated and checked against the input threshold F and if this is greater than F, it is included in the set T_k which contains the discovered size-k trees. In the next substep, graph G is partitioned into a number of subgraphs such that each subgraph embeds a size-k tree. In the last substep, size-k subgraphs with k-1 edges are formed for each subtree t in T_k and each tree t is joined with each of these subgraphs to generate size-k subgraphs which have k edges.

The Markov chain algorithm is used to generate random networks and the frequencies of the detected subgraphs in the first step are computed in these random networks. Finally, comparing the frequencies of these subgraphs in the input PPI network and in the random networks, whether the subgraphs are motifs or not are determined. The second and third steps of the algorithm are commonly employed in motif search algorithms as we saw in Section 9.4. NeMoFinder was implemented in C++ and compared with *mfinder*, sampled *mfinder* (See sections 9.4.1.1 and 9.4.2.1) and FPF [36] algorithms and found that it has much better runtime than others and can find all of the motifs [5]. In terms of the maximum size of motifs discovered and the total number of motifs discovered, tests showed that NeMoFinder was able to find motifs up to size 12 while FPF, *mfinder* with sampling and *mfinder* reached the sizes 9, 8 and 5 respectively. The total number of motifs discovered by NeMoFinder was about ten times more than the nearest FPF algorithm.

10.5   Network Alignment

Graph alignment is the process of discovering similarity between given graphs. In the graph isomorphism problem, we tried to determine whether two graphs that looked different were in fact the same graph. In the graph alignment, however, certain differences such as added or deleted vertices between the graphs are allowed. The biological systems such as the PPI networks are often noisy, making it very difficult to find isomorphic graphs. But more than that, we would be interested to compare different graphs representing different organisms. The similarities between these different graphs of the species provide us with valuable information about the fundamental structures preserved across the species and the role of these structures in the overall function of the organism.

The *sequence alignment* problem aims to find similarities between two DNA sequences and tries to minimize the number of operations which can be insertion, deletion or substitution of an element in the sequence. Alignment of two sequences provides us information about their common origin (ancestry). In the following, sequences S_2 and S_3 have evolved from S_1 by insertion, deletion and substitution operations.

$$S_1 = ACGGCTATTAAG$$

$$S_2 = ACG - CTATT - TAG$$

$$S_3 = TACGG - T - TTAAG$$

BLAST is an efficient heuristic algorithm that performs sequence alignment [2]. Just as sequence alignment discovers similar sequences between two or more DNA

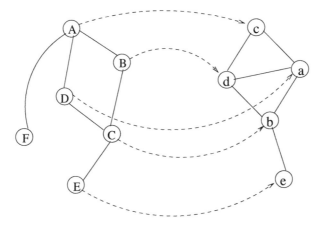

Figure 10.9: Network alignment

sequences, graph alignment aims to detect subgraphs conserved in two or more graphs representing organisms. The graph alignment problem is also closely related to the subgraph isomorphism problem as the network motif search and is NP-hard which necessitates the use of heuristics. Graph alignment in biological networks is usually called the *network alignment*. Figure 10.9 displays alignment between two networks where deletion of an edge (A, F) and addition of an edge (d, a) between the two graphs are shown.

We can formally define graph alignment as follows [7]:

Definition 16 graph alignment *Given two undirected, unweighted graphs without multiple edges and loops, let f be the bijective function $f : V_1 \rightarrow V_2$ and $Q(G_1, G_2, f)$ be the quality of alignment dependent on f. Graph alignment is finding the function f that maximizes Q.*

The bijective function f maps each vertex of G_1 to exactly one vertex of G_2 and for the alignment of graphs with different sizes, we can assume it is total and injective. Network alignment of biological networks can be classified as follows [30]:

■ *Local or Global Alignment*: Mappings between small subnetworks of two or more networks are searched in local alignment whereas each node of the first graph is aligned to exactly one node in the second graph in global alignment. PathBLAST [17], NetworkBLAST [34], NetAlign [16] and MaWIsh [15] are examples of algorithms that provide local alignment. IsoRAnk, Graemlin, GRAAL [21], H-GRAAl [25], MI-GRAAl [22] and C-GRAAL [27] are the examples of global alignment algorithms.

■ *Pairwise or Multiple Alignment*: Two networks are aligned in pairwise alignment and more than two networks are aligned in multiple alignment.

■ *Functional or Topological Information*: Functional information is the information other than the topology of the network, such as the properties of the nodes and topological information is related to connectivity of the network

10.5.1 Quality of the Alignment

The quality of the alignment can be assessed using two different approaches. While searching for network alignment between two graphs G_1 and G_2, we attempt to discover subgraphs $G_x \subset G_1$ and $G_y \subset G_2$ that are similar to each other. The alignment quality Q is clearly dependent on the number of vertices and edges that match in G_x and G_y. This way of evaluating quality of the alignment is based on graph topology and we are interested in finding the *topological similarity* between the two graphs in this case. In real networks such as PPI networks, the nodes themselves may be more or less similar to each other in which case we try to align nodes with high similarity using *node similarity*. Quality of the alignment can therefore be defined in terms of these two similarity concepts as topological quality and node similarity.

10.5.1.1 Topological Similarity

The topological similarity or the *topological quality* shows how well the structure of G_1 is preserved when mapped to G_2 under the mapping function f [29]. A standard and common metric to assess topological similarity is *edge correctness* which shows the percentage of edges from G_1 that are aligned to the edges of G_2. It can be defined formally as follows:

Definition 17 edge correctness *Given two simple graphs G_1 and G_2, Edge Correctness (EC) is:*

$$EC(G_1, G_2, f) = \frac{|f(E_1) \cap E_2|}{|E_1|} \quad (10.3)$$

It is basically a measure showing the percentage of edges in G_1 that are aligned to edges in G_2. Edge correctness value for the example graph of Figure 10.9 is 0.8 as four out of five edges of the first graph are aligned to edges in the second graph. Another measure to compare topological similarity of two or more networks is to find the size of the common connected subgraphs (CCCs) which may or may not be induced.

10.5.1.2 Node Similarity

The node similarity measure assesses the similarity between the nodes and is especially useful in PPI networks where each node represents a protein. The similarity of the protein nodes can be evaluated by testing their sequences and the BLAST algorithm is commonly used for this purpose. Node similarity is also a function of topological properties such as the degrees of the nodes or their local clustering coefficients. A function that gives different weights to topological and node similarities is usually employed in network alignment algorithms. Aligning two graphs only on node similarity can be performed by the Hungarian algorithm [23] in $O(n^3)$ time.

The *biological similarity* (or quality) considers node similarity when applied to biological networks such as the PPI networks. It also considers the correlation between the topological similarity and the node similarity (sequence alignment) and whether the alignment detects evolutionary conserved functional modules. Lastly, it would be meaningful to assess the statistical significance of the alignment performed by comparing it with the random alignment of two networks and comparing it also with the alignment between random graphs of the same sizes [30].

10.5.2 Algorithms

A number of algorithms have been developed to address the problem of network alignment. These algorithms consider local or global alignment, pairwise or multiple alignment, and topological or node similarity or a combination of both.

10.5.2.1 PathBLAST

The PathBLAST is a local network alignment and search tool that compares PPI networks across species to identify protein pathways and complexes that have been conserved by evolution [17]. It uses a heuristic algorithm with a scoring function related to the probability of the existence of a path. It searches for high-scoring alignments between pairs of PPI paths by combining the protein sequence similarity information with the network topology to detect paths of high scores. The query path input to PathBLAST consists of a sequence of several proteins, and similar pathways in the target network are searched by pairing these proteins with putative orthologs occurring in the same order in the second path. The pathways are combined to form a global alignment graph with either direct links, gaps or mismatches.

The goal of PathBLAST is to identify conserved pathways which can be performed efficiently by dynamic programming for directed acyclic graphs. However, the PPI networks are not directed or acyclic and for this reason, the algorithm employed eliminates cycles by imposing random ordering of the vertices and then performing dynamic programming which is repeated a number of times. The time complexity of the algorithm is $O(L!n)$ to find conserved paths of length L in a network of order n. Its main disadvantage is being computationally expensive and also the search is restricted to specific topology.

PathBLAST can be accessed at http://www.pathblast.org/ where the user can specify a protein interaction path as a query and select a target PPI network from the database. The output from the PathBLAST is a ranked list of paths that match the query in the target PPI network. It also provides a graphical view of the paths.

10.5.2.2 MaWIsh

Maximum Weight Induced Subgraph (MaWIsh) was proposed by Koyuturk et al. for pairwise local alignment of PPI networks [15]. It uses a mathematical model to extend the concepts of match, mismatch, and gap in sequence alignment to those of match, mismatch, and duplication in network alignment. The scoring function to rank the similarities between the graphs accounts for evolutionary events. The similarity score is based on protein sequence similarity and is calculated by BLAST.

MaWIsh attempts to identify conserved multi-protein complexes by searching for clique-like structures. These structures are expected to contain at least one hub node with a high degree. Having found a hub in a clique-like structure, it greedily extends the subgraphs. The time complexity of MaWIsh is $O(n_1 n_2)$ where n_1 and n_2 are the sizes of the two graphs being compared. The main disadvantage of MaWIsh is that it looks for a specific topological property (cliques). MaWIsh was implemented to find network alignments of the PPI networks of yeast, fly and the worm successfully.

10.5.2.3   IsoRank

IsoRank was introduced in 2008 by Singh et al. for global alignment of multiple PPI networks [32]. It is the first algorithm that provides global alignment for this purpose. Given two graphs $G_1(V_1, E_1)$ and $G_2(V_2, E_2)$ representing two PPI networks, the idea of *IsoRank* is that a protein i in G_1 is a good match for a protein j in G_2 if neighbors of i are good matches for the neighbors of j. This method is formulated as an eigenvalue problem for each pair of input networks and then k-partite matching is used to extract the final global alignment for all of the inputs. Using *IsoRank*, the global alignments for the PPI networks of yeast, fly, worm, mouse and humans were computed.

The similarity scores R_{ij} for a pair of proteins $i \in V_1$ and $j \in V_2$ is computed using the page rank algorithm we will see in Section 12.4. Then, the global alignment is computed using a greedy algorithm. It uses sequence similarity and network connectivity with other nodes to define the ranks. Pairwise similarity score R_{ij} is defined as follows:

$$R = \sum R_{ij} = \sum_{i \in N(j)} \sum_{j \in N(i)} \frac{1}{|N(i)||N(j)|} R_{ij} \qquad i \in V_i, j \in V_2 \qquad (10.4)$$

where $N(x)$ is the neighbor set of the node x. In matrix notation, Eqn. (10.4) can be rewritten as:

$$R = AR \qquad (10.5)$$

The parameter α introduced at this point defines the level of sequence similarity to be incorporated. When $\alpha = 0$, only sequence information is used and when $\alpha = 1$, the topological information is solely used. Rewriting Eqn. (10.4) using α yields:

$$R = \alpha AR + (1 - \alpha) \qquad 0 \le \alpha \le 1 \qquad (10.6)$$

or

$$R = (\alpha A + (1 - \alpha) E 1^T) R \qquad (10.7)$$

Eqn. (10.7) can be solved for R by finding the eigenvector corresponding to the eigenvector consisting of 1s. After determining the R vector, either one-to-one or many-to-many mapping can be performed. In one-to-one mapping, any node in G_1 is mapped to at most one node in G_2 and clusters of orthologous genes from different networks are discovered in many-to-many mapping. The mapping criteria of the *IsoRank* algorithm is to detect pairs of nodes with high R_{ij} scores which obey transitivity closure property, that is, if (i,j) and (j,k) are elements of mapping, then (i,k) is also included in mapping. In pairwise mapping, the highest scoring element is selected using the greedy method.

10.5.2.4 GRAAL

Graph aligner (GRAAL) is a global alignment algorithm based on topological similarity only [21]. Having two graphs $G_1(V_1,E_1)$ and $G_2(V_2,E_2)$ representing two PPI networks, it produces a set of ordered pairs (u,v) with $u \in V_1$ and $v \in V_2$, by matching them using the *graphlet degree signature similarity*. Graphlet degree signatures are computed for each node in each graph by finding the number of graphlets up to size 4 and then assigning a score that reflects this number. The scores of the nodes in the graphs are then compared to find similarities.

More specifically, GRAAL first selects a single seed pair of nodes with high graphlet degree signature similarity and then expands the alignment radially around the seed using a greedy algorithm. It first computes cost of aligning each node of G_1 with each node of G_2 by considering graphlet degree signature similarities between them. The cost of aligning two nodes u and v is computed as follows:

$$C_{uv} = 2 - (1-\alpha)\frac{\delta_u + \delta_v}{\Delta(G_1) + \Delta(G_2)} + \alpha S_{uv} \qquad (10.8)$$

where $\alpha \in [0,1]$ is a parameter that regulates the contribution of node degrees to the cost function, δ_x is the degree of the node x, Δ is the maximum degree of a graph and S_{uv} is the graphlet degree similarity of the nodes u and v. Having selected nodes u and v as the seed of a sphere of radius r around them, the algorithm then searches pairs (u',v') where $u' \in S_{G_1}$ and $v' \in S_{G_2}$ within this radius, which can be aligned with minimum cost. These nodes should not have been aligned previously. The algorithm continues until each node of G_1 is aligned with exactly one node of G_2.

H-GRAAL was proposed for better alignment at the cost of increased computation complexity [25]. The assignment in *H-GRAAL* is achieved by the Hungarian algorithm. The more recently introduced *MI-GRAAL* (Matching Integrative *GRAAL*) uses graphlet degree signature similarity, local clustering coefficient differences, degree differences, eccentricity similarity and node similarity based on BLAST [22].

C-GRAAL (common neighbors-based GRAAL) algorithm [27] uses topological similarity only and is based on the idea that the neighbors of the mapped nodes in graphs $G_1(V_1,E_1)$ and $G_2(V_2,E_2)$ should have mapped neighbors. The *node density* nd of a node v in C-GRAAL is defined as the sum of the degrees of the nodes in its closed neighborhood as follows:

$$nd_v = \sum_{u \in N[v]} \delta_u \qquad (10.9)$$

The *combined neighborhood density cnd* based on node density for nodes $u \in V_1$ and $v \in V_2$ is defined as follows:

$$cnd_{uv} = \frac{nd_u + nd_v}{max_{w \in V_1}(nd_w) + max_{y \in V_2}(nd_y)} \qquad (10.10)$$

where the denominator is the sum of the maximum neighborhood densities in each network. C-GRAAL works in three steps. It first selects the pair (u, v), $u \in V_1$ and $v \in V_2$ with the highest *cnd* value as the seed and expands the subgraphs around this seed by greedily aligning their direct neighbors. It then aligns common neighbors of already aligned nodes and repeats these two steps while there are aligned pairs of nodes that have at least one unaligned neighbor. In the final step, it greedily aligns all of the unaligned nodes in G_1 to nodes in G_2 based on the node similarity only [27]. The computational complexity of C-GRAAL is $O(n_1 n_2 + max(m_1, m_2))$ where n_1 and n_2 are the orders and m_1 and m_2 are the sizes of the graphs G_1 and G_2 respectively.

10.5.2.5 Recent Algorithms

A non-linear integer algorithm was proposed by Klau [20] for mapping in the first step. It was then shown how to linearize this problem and a Lagrangian relaxation method was given as the final step. El-Kebir et al. improved the method of Klau by modifying the upper and lower bounds of the relaxation in the tool developed called Natalie [9] which is a tool for pairwise global network alignment.

The GHOST proposed by Patro and Kingsford [29] is a pairwise global network alignment algorithm that employs spectral signatures to evaluate topological similarity between networks. The spectral signature for a node is based on the normalized Laplacian for subgraphs of various radii centered around that node. It then uses seed-and-extend strategy and an iterative local search to enlarge the graph around nodes.

SPINAL (scalable protein interaction network alignment) was developed by Aladag and Erten [1] and consists of two phases. The first phase is the coarse-grained alignment phase where all pairwise similarity scores based on pairwise local neighborhood matchings are determined. Using the produced similarity scores, the final one-to-one mapping is output by the fine-grained alignment phase by iteratively growing a locally improved solution subset that uses the similarity scores. The neighborhood bipartite graphs and the contributors are constructed in these two phases. Aladag and Erten evaluated the performance of SPINAL in the PPI networks of yeast, fly, worm and human and showed that it is scalable and outperforms various current algorithms.

10.6 Chapter Notes

Proteins have vital functions for life and they interact with each other forming PPI networks. Modeling a PPI network by a graph and analyzing the topological properties of this graph provide us with important information about the functioning of the PPI network from which health and disease states of an organism may be predicted. PPI networks are scale-free and hence exhibit few high degree nodes called hubs and many low degree nodes. They are small-world networks with relatively low diameters when compared with their sizes. These networks also have high clustering coefficients and are structured hierarchically.

Regions of PPI networks with dense interactions are called protein complexes and these structures usually have an attributed functionality. Therefore, discovering the complexes in a PPI network provides us with valuable information about the functioning and processes within that network. We have described representative algorithms to detect complexes in PPI networks in this chapter. Although the graph clustering algorithms we have discussed in Chapter 8 may also be used for this purpose, the algorithms described in this chapter in general, aim at finding complexes in PPI networks, and they have also been experimented in these networks in various studies. Finding network motifs in PPI networks is another fundamental problem to be addressed and we have briefly described the specific problems that are encountered in PPI networks while searching for motifs in these networks. The exact algorithms are time consuming in many cases and probabilistic algorithms that evaluate motif existence in subgraphs are usually preferred due to the gain in computation time.

Network alignment is the process of comparing two or more biological networks such as the PPI networks of the same kind of organism and evaluates the similarity between these networks. By discovering preserved structures across species, it is possible to detect phylogenetic relationships between them. We have briefly defined the network alignment problem, discussed the quality of the alignment and surveyed a few common algorithms and tools used for this problem, starting from the ones reported earlier and concluding with the very recently proposed algorithms. The algorithms are typically classified on their ability to perform pairwise or multiple alignment, and local and global alignment.

In terms of algorithmic challenges, discovering protein complexes, finding network motifs and testing network alignment in PPI networks are computationally difficult tasks and remain as fundamental problems. There are effective sequential heuristic algorithms for these problems as we have outlined but there is a lack of parallel and distributed approximation algorithms to solve these time consuming tasks and we believe this is a potential research area possibly with many promising results.

Exercises

1. Discuss the importance of discovering protein complexes in PPI networks. What functions can be attributed to these complexes?

2. Describe the topological properties of PPI networks in terms of the network model, clustering coefficient and the diameter of these networks by giving reasons.

3. Show the execution of *HCS* algorithm in the sample graph of Figure 10.10 to find the highly connected subgraphs of this graph.

4. Work out the k-core and core clustering coefficient values for each node of the graph in Figure 10.11. Find the weights for each node using the MCODE algorithm. What do these weights represent?

5. Find the edge correctness value for the two graphs shown in Figure 10.12.

6. Compare the network alignment algorithms PathBLAST, MaWish, IsoRank and GRAAL in terms of algorithms employed and time complexities.

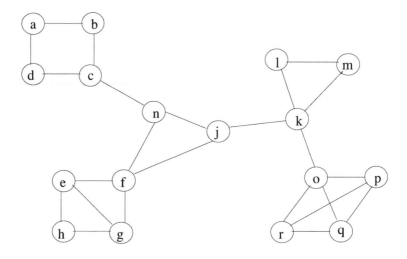

Figure 10.10: Example graph for Ex. 3

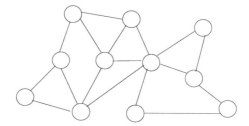

Figure 10.11: Example graph for Ex. 4

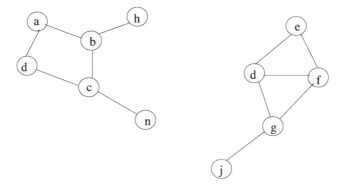

Figure 10.12: Example graph for Ex. 5

References

[1] A.E. Aladag and C. Erten. SPINAL: scalable protein interaction network alignment. *Bioinformatics*, 29(7):917-924, 2013.

[2] S. Altschul, W. Gish, W. Miller, E. Myers, and D. Lipman. Basic local alignment search tool. *Journal of Molecular Biology*, 215(3):403-410, 1990.

[3] G.D. Bader and C.W.V. Hogue. An automated method for finding molecular complexes in large protein interaction networks, *BMC Bioinformatics*, 4(2), 2003.

[4] S. Brohee and J. van Helden. Evaluation of clustering algorithms for protein-protein interaction networks. *BMC bioinformatics*, 7:488, 2006.

[5] J. Chen, W. Hsu, M.L. Lee, and S-K Ng. NeMoFinder: dissecting genome-wide protein-protein interactions with meso-scale network motifs. In *Proceedings of the 12th ACM SIGKDD International Conference on Knowledge Discovery and Data Mining*. Pages 106-115, ACM New York, 2006.

[6] M.C. Constanzo et. al. YPD, PompelPD and WorkPD: Model organism volumes of the BioKnowledge library and integrated resources for protein information. *Nucleic Acids Research*, 29:75-79, 2001.

[7] C. Doepmann. Survey on the Graph Alignment Problem and a Benchmark of Suitable Algorithms. BS thesis. Humboldt Universitat Zu Berlin, Institut fur Informatik, 2013.

[8] S.V. Dongen. Graph Clustering by Flow Simulation. PhD Thesis, University of Utrecht, The Netherlands, 2000.

[9] M. El-Kebir, J. Heringa, and G.W. Klau. Lagrangian relaxation applied to sparse global network alignment. In *Proceedings of the 6th IAPR International Conference on Pattern Recognition in Bioinformatics*, Springer, pages 225-236, 2011.

[10] F. Glover. Future paths for integer programming and links to artificial intelligence. *Computers and Operations Research*, 13(5):533-549, 1986.

[11] E. Hartuv and and R. Shamir. A clustering algorithm based on graph connectivity. *Information Processing Letters*, 76(4):175-181, 2000.

[12] Hermjakob et. al. The hupo psi's molecular interaction format: a community standard for the representation of protein interaction data. *Nat. Biotechnol*, 22(2):177-183, 2004.

[13] H. Jeong. The large scale organization of metabolic networks. *Nature*, 407(6804): 651-654, 2000.

[14] P. Jonsson and P. Bates. Global topological features of cancer proteins in the human interactome. *Bioinformatics*, 22(18):2291-2297, 2006.

[15] M. Koyuturk, Y. Kim, U. Topkara, S. Subramaniam, W. Szpankowski, and A. Grama. Pairwise alignment of protein interaction networks. *Journal of Computational Biology*, 13(2):182-199, 2006.

[16] Z. Liang, M. Xu, M. Teng and L. Niu. Comparison of protein interaction networks reveals species conservation and divergence. *BMC Bioinformatics*, 7(1):457, 2006.

[17] B.P. Kelley, R. Sharan, R.M. Karp, T. Sittler, D.E. Root, B.R. Stockwell, and T. Ideker. Conserved pathways within bacteria and yeast as revealed by global protein network alignment. In the *Proceedings of PNAS 2003*, Volume 100(20), pages 11394-11399, 2003.

[18] B. W. Kernighan and S. Lin. An efficient heuristic procedure for partitioning graphs, *The Bell System Technical Journal*, 49(2):291-307, 1970.

[19] A. D. King, N. Przulj and I. Jurisica, Protein complex prediction via cost-based clustering, *Bioinformatics*, 20(17): 3013-3020, 2004.

[20] G.W. Klau. A new graph-based method for pairwise global network alignment. *BMC Bioinformatics*, 10(Suppl 1):S59, 2009.

[21] O. Kuchaiev, T. Milenkovic, V. Memisevic, W. Hayes, and Natasa Przulj. Topological network alignment uncovers biological function and phylogeny. *Journal of the Royal Society Interface*, 7(50):1341-1354, 2010.

[22] O. Kuchaiev and N. Przulj. Integrative network alignment reveals large regions of global network similarity in yeast and human. *Bioinformatics*, 27(10):1390-1396, 2011.

[23] H.W. Kuhn. The Hungarian method for the assignment problem. *Naval Research Logistic Quarterly*, 2:83-97, 1955.

[24] M. Matula. Determining edge connectivity in O(nm) time. In the *Proceedings of 28th IEEE Symp. on Foundations of Computer Science*, pages 249-251, 1987.

[25] T. Milenkovic et al. Optimal network alignment with graphlet degree vectors. *Cancer Informatics* 9:121-137, 2010.

[26] H. W. Mewews et. al. MIPS: A database for genomes and protein sequences. *Nucleic Acids Research*, 30(1):31-34, 2002.

[27] V. Memievic and N. Pruzlj. C-GRAAL: Common-neighbors-based global GRaph ALignment of biological networks. *Integrative Biology* 4(7):734-743, 2012.

[28] M.E.J. Newman. The structure and function of complex networks. *SIAM Review*, 45(2):167-256, 2003.

[29] R. Patro and C. Kingsford. Global network alignment using multiscale spectral signatures. *Bioinformatics*, 28(23):3105-3114, 2012.

[30] N. Przulj. Graph theory analysis of protein-protein interactions. In *Knowledge Discovery in Proteomics*, edited by Igor Jurisica and Dennis Wigle, CRC Press, 2005.

[31] R. Sharan, S. Suthram, R. M. Kelley, T. Kuhn, S. McCuine, P. Uetz, T. Sittler, R. M. Karp and T. Ideker. Conserved patterns of protein interaction in multiple species. *PNAS*, 102:1974-1979, 2005.

[32] R. Singh, J. Xu, and B. Berger. Global alignment of multiple protein interaction networks with application to functional orthology detection. *PNAS*, 105(35):12763-12768, 2008.

[33] S. Skiena. *Implementing Discrete Mathematics: Combinatorics and Graph Theory with Mathematica*, p. 231. Reading, MA: Addison-Wesley, 1990.

[34] R. Sharan and T. Ideker T. Modeling cellular machinery through biological network comparison. *Nature Biotechnology*, 24(4):427-433, 2006.

[35] S. Shen-Orr, R. Milo, S. Mangan and U. Alon. Network motifs in the transcriptional regulation network of *Escherichia coli*. *Nature Genetics*, 31:64-68, 2002.

[36] V. Spirin and L.A. Mirny. Protein complexes and functional modules in molecular networks. *PNAS*, 100(21):12123-12128, 2003.

[37] M. Stoer and F. Wagner. A simple min-cut algorithm. *Journal of the ACM*, 44(4):585-591, 1997.

[38] B. Titz, S.V. Rajagopala, J. Goll, R. Hauser R, M.T. McKevitt, T. Palzkill, and P. Uetz. The binary protein interactome of *Treponema pallidum*, the syphilis spirochete. *PLoS ONE* 3(5): e2292, 2008.

[39] Xenarios et. al. DIP, the database of interacting proteins: a research tool for studying cellular networks of protein interactions. *Nucleic Acids Research*, 30(1):303-5, 2002.

[40] J. Vlasblom and S. J. Wodak. Markov clustering versus affinity propagation for the partitioning of protein interaction graphs. *BMC Bioinformatics*, 10:99, 2009.

Chapter 11

Social Networks

11.1 Introduction

Social networks consist of persons or groups of people which have some kind of relationship. A social network can be modeled by a graph where a vertex represents a person or a group and an edge between two vertices indicates an interaction between these two entities. We have already discussed the small-world property observed in many social networks in which the nodes of the network can access other nodes using a few links.

We will now take a closer look at the internal structures of social networks and investigate their properties that are radically different from those of other complex networks. As a first attempt, we will try to label the relationship between people as positive and negative and will search to assess global stability criteria for the social networks based on the stability of local relationships. The notion of *equivalence* is another property which is not found in other types of complex networks and we describe this in Section 11.3. Clustering, now called *community detection* in social networks, provides us with useful information about the structure of the network and is again a fundamental area of research in these networks. We describe four algorithms in chronological order that may be used for community detection in social networks. The first algorithms uses edge betweenness of vertices to divide the network into clusters. The second algorithm uses the resistor model of the network and the third similar algorithm is based on random walks in the network. The last algorithm uses a new measure called *modularity* to perform clustering and validates this parameter while performing clustering.

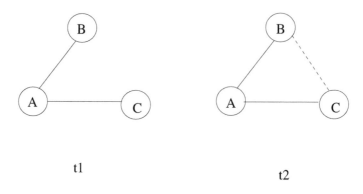

tl

t2

Figure 11.1: Triadic closure example

11.2 Relationships

In our investigation of complex networks so far, we have concentrated on static properties which we assumed do not change significantly over time. Social networks have the tendency to have a dynamic topology as can be observed in friendship networks. For example, let us assume a person A has two friends B and C at time t_1 who do not know each other. However, it is highly probable that B and C will have a chance to get acquainted in future and become friends as they have a common friend, and also they will trust each other through A. This situation is depicted in Figure 11.1 where the link between B and C is formed at time t_2. This property of social networks is known as *triadic closure* since the link between B and C closes.

11.2.1 Homophily

While social networks dynamically evolve, their formation usually follows a simple principle, the entities of the network are more likely to form relationship with entities of similar structure to themselves. For example, students at a high school will prefer to form friendship with students close to their age. This principle is known as *homophily* and is a widely investigated topic in social networks. Given a social network with two distinct populations such as boys and girls in a high school, we will search ways of evaluating if there is some homophily in such a network. In other words, do girls tend to become friends among themselves or with the opposite gender? We can evaluate the degree of homophily as in [7] where we will assume the percentage of the first population (male) is p_1 and the second one (female) is p_2. When each node of the graph modeling the social network is independently assigned the gender male with probability p_1 and female with probability p_2, the probability of having two males at the end of an edge is p_1^2 and female is p_2^2. The probability of having a male and a female at the end of an edge is $2p_1p_2$ considering both directions. We can now count the number of edges in the cutset between the two communities to find its ratio to the number of all edges and check whether this value

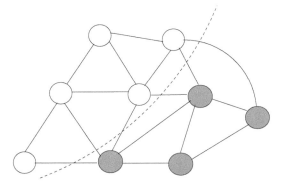

Figure 11.2: Friendship relation in a small class

is significantly lower than the expected value of $2p_1p_2$. Figure 11.2 displays the relationship between the boys and girls in a small virtual class where boys are shown white and girls are shown gray. There are 5 boys and 4 girls, giving $2p_1p_2 = 0.40$. The fraction of edges in the cutset to the total number of edges is $5/16 = 0.31$ which is significantly lower than 0.48, therefore, we can assume there is some homophily in this small class example. Specifying how much lower than the expected value can be considered as a sign of homophily is another issue to be addressed.

11.2.2 Positive and Negative Relations

Assuming the friendship is symmetric, which in fact may not always be valid, we can label an edge between two persons as positive (+) meaning they are friends or negative (-) meaning they dislike each other. The four possibilities of labeling a relationship between three people shown in Figure 11.3 are as follows:

■ The three people labeled as *A*, *B* and *C* are mutual friends as shown in (a) which is a stable and balanced configuration.

■ A single positive and two negative relations as shown in (b). This is again a balanced condition as two persons (*A* and *B*) are not friends with the third person (*C*) but they are friends with each other.

■ A person (*B*) is friends with two others (*A* and *C*) but those two friends are not friendly with each other as shown in (c) which is unbalanced as these three will not peacefully meet.

■ All of the three persons are not friendly with each other which is again an unbalanced condition.

As a result, triangles with exactly one or three positive edges are balanced, and triangles with exactly zero or two positive edges are unbalanced. We can now define

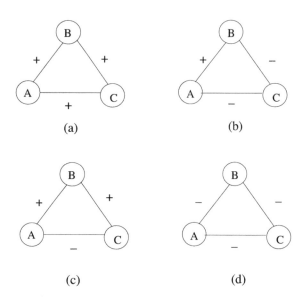

Figure 11.3: Possible friendship relations among three people

a balanced social network as the one which has all of its triangles balanced and an unbalanced social network as a network with at least one unbalanced triangle. Thus, we are attempting to assess global properties of a social network based on its local properties that can be measured relatively more easily. Figure 11.4 displays a small balanced social network in (a) and an unbalanced network in (b).

We have discussed symmetric relationships and therefore could use undirected graphs to model these relationships. However, it is common to have relationships which are asymmetric in social networks such as the "best friend" networks. This relationship clearly is not symmetric as a person *A* may think *B* is her best friend but *B* may have somebody else as her best friend. In order to model such a social network, we need to use directed graphs. In this case, we will have six different relationship between the three people in a triangle.

11.2.3   Structural Balance

Given a complete labeled graph G of a social network, let us assume that G can be divided into two groups of nodes as V_1 and V_2 where all people within V_1 and within V_2 are friends with people in their group. In order to have a balanced network, each person in each group must dislike all other people in the other group. This property called the *balance theorem* by Harary [4] can be stated formally as below.

Theorem 11.1 Balance Theorem
There are two cases for a labeled complete graph representing a social network to be balanced. Either all pairs of nodes are friends with each other; or the nodes of the

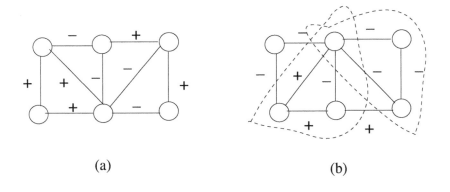

(a)   (b)

Figure 11.4: a) A balanced network. b) An unbalanced network with unbalanced triangles shown by dashed regions

network can be divided into two groups V_1 and V_2 where all nodes in V_1 are friends with each other and also all nodes in V_2 are friends with each other, and all nodes of V_1 dislike all nodes of V_2.

An example of balance theorem for the complete graph K_5 is shown in Figure 11.5 where nodes a, b and c are one group of people (V_1) who are all mutual friends and d and e are the members of the other group (V_2) and are also friends to each other. However, each member of each group is not friendly with all members of the other group as shown where the cutset between the two groups have all negative signs and this is a balanced condition since all triangles in the network are balanced.

The manifestation of the balance theorem has been observed in international relations during world wars and international crisis where nations were divided into

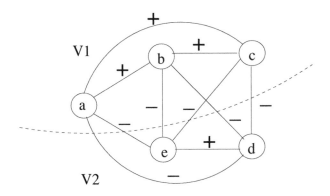

Figure 11.5: A balanced K_5 graph

two groups. All of the nations in each side were allies with each other and they were enemies with each nation in the other group. This situation was perfectly stable and probably contributed to the long duration of the wars.

11.3   Equivalence

It may sometimes be useful to investigate the position of a person or a group of persons in a social network rather than their properties. In this case, we will be searching similarity between the nodes. Three different ways of evaluating equivalence are *structural equivalence, automorphism* and *regular equivalence.*

Structural equivalence aims to find similarities between persons or a group of people in terms of the likeliness of their neighbors [36]. In an undirected graph, two nodes u and v are structurally equivalent if they have the same set of neighbors. In a directed graph, the in-neighbors and out-neighbors of u and v should be the same for structural equivalence [6]. In the undirected graph of Figure 11.6, the two nodes u and v are structurally equivalent as they have the same set of neighbors ($\{a,b,c\}$).

If two nodes that are structurally equivalent change their positions in the network, we expect no change in the operation of the network. For example, if there are two functionally equivalent surgeons who can operate with the same group of nurses, operations can be performed comfortably by either of them. However, we may be interested to find persons of similar positions rather than exact positions in a network. We may be looking for nurses in a hospital for example. In essence, we are trying to exchange two nodes along with their neighbors so that the resulting graph remains the same. This problem is known as *automorphism* and is one-to-one mapping from the vertices of a graph G to its vertices such that edge-vertex connectivity is preserved. Now we can define automorphic equivalence as follows [11]:

Definition 18 *Two vertices u and v of a graph G(V,E) are automorphically equivalent if there is an automorphism of G which maps u to v.*

The regular equivalence is defined recursively with respect to the positions or roles of the nodes in a social network [3]. If two nodes u and v have the same position or role in a social network, they are regularly equivalent. In order to assess whether

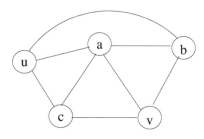

Figure 11.6: Structurally equivalent nodes

u and v have the same role, we need to check their neighbors. If their neighbors are also regularly equivalent, then u and v are regularly equivalent.

11.4 Community Detection Algorithms

The graph clustering algorithms we studied in Chapter 8 can be used to detect communities in social networks which are represented by graphs. Our aim in this section is to introduce algorithms that have been experimented and used in social networks. We will describe four such algorithms based on edge betweenness and modularity properties of graphs.

11.4.1 Edge Betweenness-based Algorithm

The betweenness centrality of a vertex was defined in Section 4.6 as the ratio of the number of shortest paths that pass through a vertex to the number of all shortest paths in the graph. The edge betweenness of an edge e was also described as the ratio of the number of shortest paths between any vertex pairs that pass through e to total number of shortest paths. If there are two or more shortest paths between two vertices, equal weights are assigned to all edges so that the total weight is unity.

The community detection algorithm proposed by Girvan and Newman [2] is based on the idea that an edge e_i which has a high edge betweenness value has a higher probability of joining two communities than an edge e_j which has a lower value. Thus, this algorithm removes the edges with the highest edge betweenness values at each iteration with the expectation of separating communities at some point during this process. It is a divisive algorithm since it starts with a single cluster and iteratively divides the clusters into a number of clusters as shown in Alg. 11.1.

Algorithm 11.1 *GN_Alg*

1: **Input** : Undirected, unweighted graph $G(V,E)$
2: **Output** : Clusters $C \leftarrow \{C_1, C_2, ..., C_m\}$
3: $G_w(E_w, V_w) \leftarrow G(V,E)$ ▷ initialize
4: **repeat**
5:     $G'_w(E'_w, V'_w) \leftarrow Between\_Alg(G_w(E_w, V_w))$ ▷ calculate betweenness values
6:     **find** the heaviest edge $e_{uv} \in E'_w$
7:     $E'_w \leftarrow E'_w \setminus \{e_{uv}\}$ ▷ remove heaviest edge
8:     $G_w(E_w, V_w) \leftarrow G'_w(E'_w, V'_w)$ ▷ update
9: **until** required number of clusters are formed

Calculation of Edge Betweenness

In the simplest case, given an unweighted, undirected graph $G(V,E)$, the BFS algorithm is run to find the shortest paths from a single vertex in $O(n+m)$ time to obtain a BFS tree T. Let us assume all of the shortest paths found by BFS algorithm are

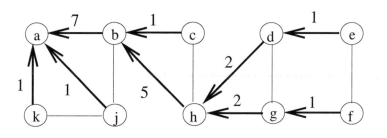

Figure 11.7: Edge betweenness values for unique shortest paths

unique, which means there is only a single shortest path between any pairs of vertices. We can then identify the leaf vertices of T which do not have any children, and a weight of unity is assigned to every edge that connects these leaf vertices to the rest of the tree. Afterwards, moving upwards, each edge e_i is assigned a weight that is one more than the sum of the weights of the edges below it to account for the shortest paths that run through e_i towards the vertex a. This process is repeated for each of the n vertices of the graph G to obtain edge betweenness values for all edges in G. Since assigning weights to edges also takes $O(m)$ time for each source vertex, total time of this algorithm is $O(nm)$. Figure 11.7 displays a sample graph with ten vertices $a, .., k$. The BFS tree shown by bold arrows consists of unique paths and the edges are labeled with the betweenness values for this BFS tree, starting with the leaf vertices e and f.

However, there are many cases of more than one shortest paths between pairs of vertices in a graph. In these cases, the importance of an edge decreases in inverse proportion to the number of shortest paths that run through it. Girvan and Newman proposed an algorithm to find the edge betweenness values for edges in a graph where there are more than a single shortest path between the pairs of vertices, in two steps. In the first step, a modified BFS algorithm for each vertex of the graph is run and for each root vertex s, every other vertex v is labeled with a weight showing the number of shortest paths from v to s. The pseudocode of the algorithm shown in Alg. 11.2, runs similar to the BFS algorithm of Section 3.4, however, a weight is now associated with vertices instead of levels. The weight of an explored vertex v is changed to the sum of the weight of its lower neighbor u and its weight, if it is at a distance of one more hops to the source vertex s than u as shown in lines 17 and 18 of the algorithm. The predecessors of a vertex can now be more than one node to account for multiple shortest paths to the source s.

In the second step, leaf vertices are identified first and for any such vertex u, any edge e_{uv} incident to u and its neighbor v is given a weight of w_u/w_v. The edge weights are then computed toward the source vertex s by assigning a weight to an edge (u, v) with v being a predecessor of u in the BFS tree, the sum of all edge weights below it plus 1, multiplied w_v/w_u. An example implementation is shown in Figure 11.8 for the source vertex a. The vertex weights are determined by the *Weigh_Vertex* algorithm

---

Algorithm 11.2 $Weigh\_Vertex(G(V,E),s)$

1: **Input** : undirected, unweighted graph $G(V,E)$
2: **Output** : vertex weighted graph $G_w(E,V_w)$
3: **for all** $u \in V \setminus \{s\}$ **do**   ▷ initialize distances
4:       $d_u \leftarrow \infty$
5:       $preds(u) \leftarrow \perp$
6: **end for**
7: $d_s \leftarrow 0$
8: $w_s \leftarrow 1$
9: $Q \leftarrow s$
10: **while** $Q \neq \emptyset$ **do**   ▷ continue until Q is empty
11:       $u \leftarrow deque(Q)$ ▷ get the first element
12:       **for all** $v \in N(u)$ **do**   ▷ check all neighbors
13:             **if** $d_v = \infty$ **then**
14:                   $d_v \leftarrow d_u + 1$
15:                   $w_v \leftarrow w_u$
16:             **else if** $d_v = d_u + 1$ **then** $w_v \leftarrow w_v + w_u$ ▷ check for multiple parents
17:                   $preds(v) \leftarrow preds(v) \cup \{u\}$
18:                   $enque(Q,v)$
19:             **end if**
20:       **end for**
21: **end while**
22: **return** $G_w(V_w,E)$

---

initially. There are three leaf vertices k, d and e, and the weights to edges (d,c), (e,c) and (e,f) are labeled initially. Adding unity to the sum of the edges that are lower than c in BFS tree and multiplying it by w_b/w_c results in 4/3 so the edge (c,b) is labeled with 4/3 and (c,g) is labeled with 4/3 accordingly. This process is repeated for all vertices and we can see that edge (a,g) has the highest betweenness value for

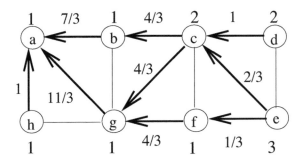

Figure 11.8: Edge betweenness values for multiple shortest paths

source vertex *a* in this graph since it has seven shortest paths running through it, from any vertex to the source *s*.

A detailed implementation of this procedure for a single source vertex *s*, which starts by calling the vertex labeling algorithm first is shown in Alg.11.3 where a weight value is associated with a vertex *v* and *preds(v)* is the set of vertices found by the *Weigh_Vertex* algorithm that are prior to *v* in the BFS tree, to account for multiple paths. The algorithm traverses the BFS tree upward by assigning weights to the edges in the described manner.

Algorithm 11.3 *Between_Alg(G(V,E), s)*

1: **Input** : undirected, vertex weighted graph $G(V,E)$
2: **Output** : vertex and edge weighted graph $G_w(E_w, V_w)$
3: $preds(v) \leftarrow$ predecessors of vertices after the *Weigh_Vertex* algorithm
4: $W = \{w_1, w_2.., w_n\}$ weights of vertices
5: $WE = \{w_{ij}\} \leftarrow 0$ weights of edges
6: $WAW = \{we_1, we_2, .., we_m\} \leftarrow 0$ accumulated edge weights on vertices
7: $G_w(V_w, E) \leftarrow Weigh\_Vertex(G(V,E), s)$
8: **for all** $u \in V_w$ and *u* is a leaf **do**   ▷ process leaves
9:     **for all** $v \in preds(u)$ **do**
10:         $w_{uv} \leftarrow (w_v/w_u)$ ▷ assign weights to preceding edges
11:         $we_v \leftarrow we_v + w_{uv}$ ▷ accumulate preceding edge weights
12:         **if** $v \notin Q \wedge v \neq s$ **then** $Enque(Q, v)$ ▷ insert *v* in *Q* if not in
13:         **end if**
14:     **end for**
15: **end for**
16: **while** $Q \neq \emptyset$ **do**   ▷ continue until *Q* is empty
17:     $u \leftarrow Deque(Q)$
18:     $w_u \leftarrow w_u + 1$
19:     **for all** $v \in preds(u)$ **do**
20:         $w_{uv} \leftarrow we_u \times (w_v/w_u)$ ▷ assign weights to preceding edges
21:         $we_v \leftarrow we_v + w_{uv}$ ▷ accumulate preceding edge weights
22:         **if** $v \notin Q \wedge v \neq s$ **then** $Enque(Q, v)$
23:         **end if**
24:     **end for**
25: **end while**
26: **return** $G_w(V_w, E_w)$

Calculation of edge betweenness is the most time consuming part of this algorithm and repeating for *n* vertices and summing the results provide edge betweenesses in $O(mn)$ steps for unweighted graphs. Total cost for all edges is $O(m^2 n)$ which makes it unsuitable for networks with more than few thousand nodes.

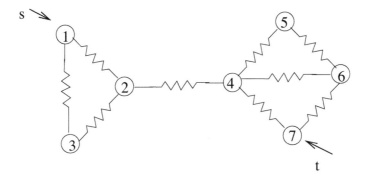

Figure 11.9: A resistor network

11.4.2 Resistor Networks

Newman et al. proposed two other methods to detect community structure in networks [9]. In the *resistor networks* method, the network is considered as an electrical circuit in which each edge is assigned a unit resistance and pair of nodes act as unit voltage sources and voltage sinks. Given a graph $G(V,E)$ representing the network, we can calculate the current through an edge $e_x \in E$ for a unit voltage source $s \in V$ and sink $t \in V$ using Kirchoff's current and voltage laws which state that the sum of currents entering a node should equal the sum of currents exiting the node and sum all of the mesh voltages in a loop shoul be equal to zero. Since current will flow among shortest paths from any source to any sink, total current through any edge will be proportional to the number of shortest paths through it. We can then proceed as in the previous algorithm by removing the edge with the highest total current through it and continue until the network is disconnected. A resistor network with nodes numbered 1,...,7 is shown in Figure 11.9. As can be seen, assigning unity voltage to a source and ground to the sink for all node pairs and summing the currents that pass through the edges results in the maximum current to flow through nodes 2 and 4, removal of which will result in a disconnected network even in the first step.

This so called *current flow betwenness* values for edges can be formally calculated as follows [9]. Given the adjacency matrix $\mathbf{A}$ of a graph G representing the network, the relation between voltage $\mathbf{V}_i$ at vertex i due to source s and sink t can be calculated as follows:

$$\sum_j A_{ij}(V_i - V_j) = \delta_{is} - \delta_{it}, \tag{11.1}$$

where A_{ij} is the ij element of matrix $\mathbf{A}$. Clearly, only the voltages of the neighboring nodes of node i will contribute to the total current flowing out of node i. In matrix notation, Eqn. (11.1) can be expressed as $(\mathbf{D} - \mathbf{A}) \cdot \mathbf{V} = s$ where $\mathbf{D}$ is the diagonal degree matrix and the source vector $s_i = 1$ when i is the source; $s_i = -1$ when i is the sink and it is 0 otherwise. Our aim is to obtain $\mathbf{V}$ to find the node with the highest current value but we cannot multiply each side of Eqn. (11.1) by $(\mathbf{D} - \mathbf{A})^{-1}$ to get the voltage vector $\mathbf{V}$ as the Laplacian matrix $(\mathbf{D} - \mathbf{A})$ is singular. We can however select any vertex v as a reference point and remove rows and columns corresponding to v from $\mathbf{D}$ and $\mathbf{A}$ before inversion. We can therefore write the new equation as:

$$\mathbf{V} = (\mathbf{D}_v - \mathbf{A}_v)^{-1} \cdot s \tag{11.2}$$

where $\mathbf{D}_v$ and $\mathbf{A}_v$ are the matrices with the rows and columns for the selected vertex v removed. The procedure to calculate the total current through an edge (i, j) is as follows [9]. We first remove any vertex v from $\mathbf{D}$ and $\mathbf{A}$ to get $\mathbf{D}_v - \mathbf{A}_v$. We then invert $\mathbf{D}_v - \mathbf{A}_v$ to find the voltage vector $\mathbf{V}$ for every pair of sink and source vertices. Then, the sum of currents through the edge (i, j) for every voltage difference between its vertices i and j will provide us the current flow betweenness value for the edge (i, j). The matrix inversion has $O(n^3)$ time complexity, and calculation of edge betweenness takes a further $O(mn^2)$ time. Total time taken therefore is $O((n+m)mn^2)$ which becomes $O(n^4)$ in sparse graphs. The performance of this algorithm is not favorable for large graphs with number of vertices higher than few hundred.

11.4.3   Random Walk Centrality

Random walk betweenness of an edge edge (i, j) is defined as the total number of expected random walks that pass through (i, j) between all vertex pairs s and t [9]. A random walk at vertex i selects uniformly a neighbor j of i. The probability of selecting a particular node j of the graph G is $A_{ij}/deg(i)$ where $\mathbf{A}$ is the adjacency matrix of the network graph and $deg(i)$ is the degree of vertex i. The matrix with these elements can be formed as $\mathbf{M} = \mathbf{A} \cdot \mathbf{D}^{-1}$ where $\mathbf{D}$ is the diagonal matrix as before. We need to find random walks that reach the destination vertex t and stop there, and this can be achieved by removing the vertex t from the graph G to prevent any walk starting from t to reach any other vertex. We can therefore state $\mathbf{M}_t = \mathbf{A}_t \cdot \mathbf{D}_t^{-1}$ where the matrices do not contain rows and columns for vertex t now. We can now find the probability of a walk that starts at vertex s and terminates at a vertex u which is different from vertex t, and takes n steps. In this case, walks that start at vertex s terminate at vertices u and v with probabilities $[\mathbf{M}_t^n]_{us}$ and $[\mathbf{M}_t^n]_{vs}$ and $1/deg(u)$ and $1/deg(v)$ of these walks pass through edge (u, v) in either direction. The average value of all random walks of any length that pass through the edge (u, v) is $deg(u)^{-1}[(\mathbf{I} - \mathbf{M}_t)^{-1}]_{us}$. In matrix notation [9]:

$$\mathbf{V} = \mathbf{D}^{-1} \cdot (\mathbf{I} - \mathbf{M}_t)^{-1} \cdot s = (\mathbf{D}_t - \mathbf{A}_t)^{-1} \cdot s \tag{11.3}$$

with source vector s having a single 1 in the position of the source being considered and 0 in all other positions. The random walk betweenness of an edge (u, v) can be specified as the absolute value of the differences of the two probabilities V_u and V_v. We had computed $(\mathbf{D}_t - \mathbf{A}_t)^{-1} \cdot s$ to find the current flow values in a similar way. There was a sink node t in the resistor network, however, we can choose the sink t to be removed from the network which means two methods are equivalent.

11.4.4 *Modularity-based Algorithm*

A general requirement to measure the performance of a clustering algorithm is to assess the quality of the clusters formed after running the algorithm. Newman introduced the *modularity* concept to quantify the strength of the clusters formed [8]. Given a graph with k modules, let us assume that e_{ii} is the percentage of edges in module i, and a_i is the percentage of edges with at least one end in module i. The modularity parameter is defined as follows [8]:

$$Q = \sum_{i=1}^{k}(e_{ii} - a_i^2) \tag{11.4}$$

The term inside the summation shows the difference in probabilities of an edge being in module i and that a random edge would fall in module i. High modularity shows that there are more edges within the module than expected. The maximum value of Q is 1 and any value approaching unity shows strong community structure in the network. For an undirected, unweighted graph G with existing k modules, let us form a symmetric matrix $M[k,k]$ with each element m_{ij} being the percentage of edges between modules i and j. Therefore, m_{ii} is equal to the e_{ii} parameter and the sum of the ith row of M except the diagonal elements, yields the a_i parameter of Eqn. (11.4).

Figure 11.10 shows a graph with four clusters $C_1,...,C_4$ and there are a total of 20 edges in this graph and the matrix M can be computed as follows

$$\begin{bmatrix} 0.15 & 0.2 & 0.05 & 0.05 \\ 0.2 & 0.05 & 0.05 & 0 \\ 0.05 & 0.05 & 0.25 & 0.05 \\ 0.05 & 0 & 0.05 & 0.15 \end{bmatrix}$$

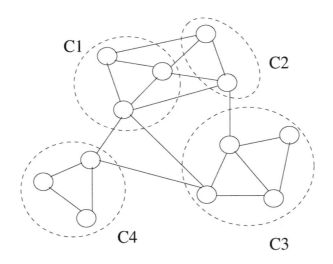

Figure 11.10: A clustered network

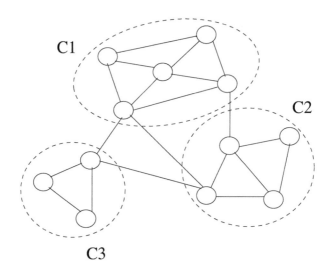

Figure 11.11: A different clustering of the network of Figure 11.10

For example, cluster C_1 and C_2 have four edges between them; therefore, $m_{1,2} = m_{2,1} = 4/20$. The diagonal elements of M are the $e_{i,i}$ values and the sum of the rows (or columns) provide the a_i values. The modularity for this graph with the existing clusters can be computed according to Eqn. (11.4) and the contribution to this value from clusters $C_1, ..., C_4$ are 0.06, -0.0125, 0.2275 and 0.14 respectively, giving a total value of 0.415 for modularity. We can visually detect that clusters C_1 and C_2 may be combined to result in a single cluster and the clusters are renamed as shown in Figure 11.11.

In this case, the matrix M becomes:

$$\begin{bmatrix} 0.4 & 0.1 & 0.05 \\ 0.1 & 0.25 & 0.05 \\ 0.05 & 0.05 & 0.15 \end{bmatrix}$$

with the first row and column representing the combined clusters C_1 and C_2. The modularity value for the graph with contributions 0.3775, 0.2275 and 0.14 from the new clusters C_1, C_2, C_3 is 0.745 which is considerably higher than the previous 0.415 value of Q. As a higher Q value indicates better formed clusters, we may decide to combine the clusters C_1 and C_2 of Figure 11.10, however, we need to consider combining each cluster pair, and perform merging the two clusters that give the highest increase in the modularity value. Girvan and Newman proposed a greedy algorithm [8] which always tries to enhance modularity based on what we have been discussing, which consists of the following steps:

1. Each node of the graph is a cluster initially.

2. Merge the two clusters that will increase the modularity by the largest amount.

3. If merges start reducing modularity, stop.

The algorithm proposed is an agglomerative hierarchical clustering algorithm as it starts with each vertex as a single cluster and combines the two clusters that increase the modularity by the largest amount into a new one. The output from this algorithm is a dendogram which can be cut by a horizontal line to obtain the required clusters. The pseudocode is similar to the agglomerative hierarchical clustering of Section 7.2 and is left as an exercise (see Ex. 5). The running time of the algorithm is $O((m+n)n)$, or $O(n^2)$ on sparse graphs [8]. Newman also proposed a method based on the spectral properties of the modularity matrix Q [10]. In this method, the eigenvector corresponding to the most positive eigenvalue of the modularity matrix is first found and the network is divided into two groups according to the signs of the elements of this vector.

11.5 Chapter Notes

We have investigated social networks in terms of their network structure in this chapter. We first looked at relationships and defined homophily which showed the tendency of similar people to have relationships between them. We provided a simple metric to have some insight about the homophily in a social network. We then defined positive and negative friendships between persons and attempted to discover the balance in the social network as a whole based on local stabilities of the three persons known as triangles. We investigated equivalence of persons in a social network and described three different ways of establishing equivalence as structural equivalence, automorphism equivalence and regular equivalence.

In order to detect community structures in social networks, we described four algorithms which may also be used for other types of complex networks. The first algorithm uses edge betweenness, the second one uses electrical circuits, the third one is based on the modularity measure and the fourth one uses spectral graph-theoretic concepts to discover clusters in social networks. The edge betweenness algorithm and its derivatives have been proposed and implemented in various complex networks. Holme et al. provided a modified version of the edge betweenness algorithm of Newman to discover sub-networks in metabolic networks [5]. The modified edge betweenness algorithm was implemented to detect clusters in gene regulatory networks [12] and the algorithm is also applied to PPI networks [1]. Modularity is a widely used measure to evaluate the quality of a clustering method. It is particularly used in social network analysis. Spectral clustering with modularity provides favorable clustering.

Although research in social networks has a long history, the research of these networks from complex networks and graph theoretical view is relatively recent. We believe there is still much research to be done in this area, especially there is a need for parallel and distributed algorithms for the computation intensive tasks involved in finding communities in social networks.

Exercises

1. Show the triadic closure relationships that may be formed in future in the example social network of Figure 11.12

2. Show whether the labeled K_4 graph in Fig 11.13 is balanced or not by checking each triangle and also by using the balance theorem. What is the cutset between the two stable groups in this graph?

3. For the example graph of Fig 11.14, work out the edge betweenness values for all edges. Then, divide this graph into three clusters using the Girvan-Newman edge betweenness algorithm.

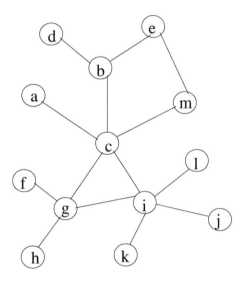

Figure 11.12: Example graph for Ex. 1

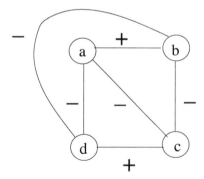

Figure 11.13: Example graph for Ex. 2

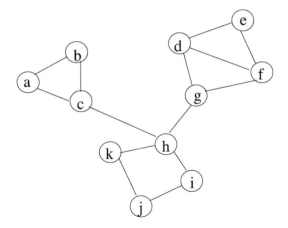

Figure 11.14: Example graph for Ex. 3

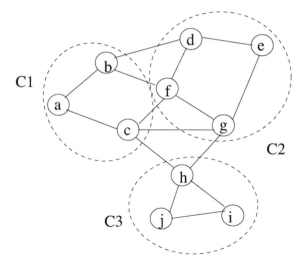

Figure 11.15: Example graph for Ex. 5

4. Write the detailed implementation pseudocode of the modularity based clustering algorithm using a similar structure to the agglomerative clustering algorithm of Section 7.3.

5. Work out the modularity values in the network of Fig 11.15.

References

[1] R. Dunn, F. Dudbridge, and C. M. Sanderson. The use of edge-betweenness clustering to investigate biological function in protein interaction networks. *BMC Bioinformatics*, 6(39), 2005.

[2] M. Girvan and M.E.J. Newman. Community structure in social and biological networks. *PNAS*, 99:7821-7826, 2002.

[3] R. Hanneman and M. Riddle. *Introduction to Social Network Methods*. On-line book. University of California Riverside, 2005.

[4] F. Harary. On the notion of balance of a signed graph. *Michigan Mathematical Journal*, 2(2):143-146, 1953.

[5] P. Holme, M. Huss, and H. Jeong. Subnetwork hierarchies of biochemical pathways. *Bioinformatics* 19(4):532-538, 2003.

[6] F. Lorrain and H. White. Structural equivalence of individuals in social networks. *Journal of Mathematical Sociology*, 1:49-80, 1971.

[7] D. Easley and J. Kleinberg. *Networks, Crowds, and Markets: Reasoning About a Highly Connected World*. Cambridge University Press, 2010.

[8] M. Newman. Fast algorithm for detecting community structure in networks. *Physical Review E 69*, 066133, 2004.

[9] M.E.J. Newman, M. Girvan. Finding and evaluating community structure in networks. *Physical Review E 69*, 026113, 2004.

[10] M.E.J. Newman. Finding community structure in networks using the eigenvectors of matrices, *Physical Review E 74*, 036104, 2006.

[11] M.V. Steen. *Graph Theory and Complex Networks: An Introduction*. Chapter 13, M.V. Steen, 2010.

[12] D.M. Wilkinson and B.A. Huberman. A method for finding communities of related genes. *PNAS*, 101, Suppl 1:5241-5248, 2004.

Chapter 12

The Internet and the Web

12.1 Introduction

Computer networks consist of computers which communicate using communication links. The Internet and the World Wide Web (Web) are the two widely used computer networks as we have seen in Chapter 1. The Internet which is the largest computer network in the world is organized hierarchically from home or office computers to networks in organizations and then to service providers which are connected via backbone networks. The Web is functionally an information network that uses the underlying Internet for data transfer.

In this chapter, we will first investigate the structure and the properties of the Internet. We will model the Internet using undirected graphs, nodes of which are routers or autonomous systems. We will then look at the results of various projects that have performed tests on the Internet using these two models and summarize the results of these tests. We will use directed graph model of the Web and find that it has a specific structure. We will describe the widely used page rank algorithm and the hubs and authorities algorithm. In both the Internet and Web analysis, we will restrict our investigation to complex network properties.

12.2 The Internet

The Internet is a worldwide computer network which connects millions of computers in the world. These computers called *hosts* may be personal computers, workstations, servers, sensors, mobile phones or various other computational devices. The hosts are physically connected to the Internet using communication links such as coaxial cable, fiberoptics or wireless medium. A *data packet* is the basic data unit transferred over the Internet and a *router* is a device in the Internet that inputs a packet from one

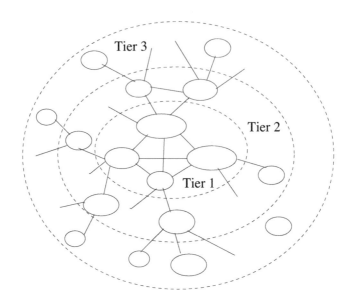

Figure 12.1: Internet tier structure

of its input ports and transfers this packet to one of its output ports. *Packet switching* is the process of transferring packets from the sending hosts to the receiving hosts using suitable paths called *routes*.

The hosts are connected to the Internet using Internet service providers (ISPs) which may provide different modes of connection to the Internet such as wireless or broadband access. The ISPs are organized hierarchically such that lower-tier ISPs are connected to the higher-tier ISPs which have more efficient routers and high-speed fiberoptic links between these routers as shown in Figure 12.1.

12.2.1 Services

We will first look at the services offered by the Internet before analyzing its structure. These services can be broadly classified as connection methods, circuit provision and the protocols offered.

12.2.1.1 Services of Connection

The hosts, routers and other devices in the Internet exchange information using *protocols* which enable correct and timely delivery of data between the users. A *protocol* is a set of rules and procedures that specify the ways of communication between the Internet devices. The Internet provides two types of services to the application as *connection-oriented* service and *connectionless* service. In a connection-oriented service, the applications that want to communicate first establish a connection between them by exchanging control messages. This is analogous to two people who first greet each other before transferring any information. In a connectionless service, the sending application simply sends its data without informing the receiver that it will communicate with it. These two modes of operation are both used in the Internet since they may be employed for different requirements. For a long lasting communication consisting of many packet transfers, the connection-oriented service which may also incorporate some reliability mechanism may be preferable. This service usually has a flow control mechanism that provides sequencing of data packets, reducing the burden of the application. However, the setting-up and termination of the connection are time costly and the connectionless service may be more preferable for time-critical applications.

12.2.1.2   Circuit and Packet Switching

In *circuit switching*, an *end-to-end connection* between the two hosts is established before the communication. As we cannot have all users connected to all users, we need to *multiplex* the communication medium among the users. Two main methods of multiplexing the communication medium are the *frequency division multiplexing* (FDM) and the *time division multiplexing* (TDM). The frequency spectrum of a communication link is shared among many applications that use a different partition of this spectrum in FDM. The *bandwidth* of the medium is the width of the allowed frequencies in that medium. In TDM, the time to use a communication medium is shared among the users.

A long message to be transferred by the sending host is usually divided into smaller packets for ease of transfer. These packets traverse network links, the routers and the link layer switches, to arrive at the receiving host. Each packet is stored at a switch or a router and then forwarded to a convenient output link of the switch or router that is on the optimal route to the destination. This process is known as *packet switching* and is the fundamental data transfer method in the Internet. Two fundamental packet switching networks are the *datagram networks* and the *virtual circuit* networks. In datagram networks, the destination address included in each packet is used to route it to the destination and a virtual connection needs to be established before communication can take place in virtual circuits. A virtual circuit identifer (VC ID) is assigned to a virtual circuit during its establishment, which is enclosed in the header of each packet transferred consequently over this virtual circuit. The path between the source and destination is determined during the virtual circuit set-up phase and all of the packets during the communication follow this path. The task of a router in this path is simply to map the VCID in the packet header to one of its outgoing links and send the incoming packet over that link [32].

12.2.1.3   Internet Protocol Suite

The Internet protocol suite consists of four layers which are called the physical layer, data link layer, network layer and the transport layer as shown in Figure 12.2. The physical layer is responsible for signal synchronization and framing at bit level whereas the data link layer is responsible for synchronization at frame level and error checking and correction. The network layer is independent from the local hardware and its main function is the routing of packets to destinations using the minimum cost

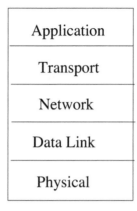

Figure 12.2: Internet protocol layers

links. The main protocol at this level is the Internet protocol (IP) which defines the fields in the datagram which is the basic data unit at this level. IP also describes how the routing protocols should use these fields. There are many routing protocols that are used at network level of the Internet which all use the structure and functionality imposed by the IP.

The transport layer provides delivery of messages between the client and server parts of the application. The two protocols at this level are the transmission control protocol (TCP) and the unreliable datagram protocol (UDP). TCP provides a connection-oriented reliable data transfer between the application processes whereas the service provided by the UDP is connectionless. TCP provides a congestion control mechanism to avoid bottleneck regions in the Internet. It breaks a long message into a number of packets for ease of delivery. TCP/IP or UDP/IP are the two main modes of data transfer methods at layers 3 and 4 of the Internet.

12.2.2 Analysis

The subnetworks of the Internet which are managed by a separate administration that uses the same procedures for routing and network management are called autonomous systems (ASs). Therefore, the Internet can be viewed as a collection of ASs connected together which means we can model the Internet as a graph with each vertex representing an AS. An alternative and a finer model of Internet can be obtained by representing it as an undirected graph of connected routers.

Several research studies have attempted to provide topological graphs (maps) of the Internet at router and AS levels. Two methods of obtaining Internet connectivity information are by inspecting the router tables and exploring by the use of software probes. The *Mercador* project provided a graph of Internet at router level and used *IP source routing* to find connectivity of the Internet from a single source router [27]. The *Oregon Route-Views* (RV) project provided the map of Internet at AS level with data from router tables stored at the nodes using the border gateway protocol (BGP) [29]. The *Distributed Internet Measurements and Simulations* (DIMES) project investigated Internet structure at both AS and router level by the use of distributed agents which collected data using various tools [16, 17]. The CAIDA project placed several monitors to probe the communication over the Internet at important points [9].

The topology of the Internet can then be investigated to find the degrees of the nodes, their shortest paths to all other nodes, their betweenness values and clustering coefficients. The findings of these projects for the properties of the Internet at AS and router levels can be summarized as follows [8]:

■ The average degree of the nodes in Internet is very small, between 2 and 8, compared to its size. The Internet connectivity graph is therefore sparse.

■ The degree distribution of the nodes in Internet show a heavy-tailed distribution which means the nodes with very high degrees are scarce and most of the nodes have low degrees. We can therefore say Internet is a *scale-free* network as various other complex we have seen.

■ The average shortest path length is very small, between 3 and 9, which means Internet is a *small-world* network.

The Internet has few high degree ASs and many low degree ASs from which we can conclude that the Internet is a scale-free network also when considered as a network of ASs. The *hop plot* of Internet is defined as the average number of nodes that are at distance of at most l from any node as follows:

$$M(l) = n \sum_{k=0}^{l} P(k) \tag{12.1}$$

where $P(k)$ is the probability distribution of finding two nodes separated by distance l [23, 8]. The value of $k = 1$ includes all of the nodes in the closed neighborhood of a node and when k is equal to the diameter of the network, we have all of the nodes covered. Data from CAIDA, DIMES, RV and Mercador projects were used to find hop plots of Internet at AS and router levels in [8]. At AS level, all of the connectivity data was used and at router level, 10^3 source routers were selected at random and shortest paths to all other routers from these source routers were calculated. The hop plots at both AS and router levels showed exponential growths of the hop plot M values with respect to the hop count l values as would be expected in a small-world network. The exponential curves for the router maps grows more sharply then the ones for AS maps.

12.3 The Web

The Web is a distributed information network which has been developed for sharing data over the Internet. Web consists of many *sites* which store Web documents called

Web pages. Each site is identified by a *domain name* such as www.izmir.edu.tr and is maintained by *Web servers.* A request to a Web server is made by a *Web client* using programs called *browsers.* Web uses the *Hyper Text Transfer Protocol* (HTTP) for communication between the Web clients and the Web servers. A Web page may reference another document using a hyperlink in the format of uniform resource locator (URL) [36]. For example, the Web page for Stanford University (www.stanford.edu) has a hyperlink to URL http://www.stanford.edu/academics.html which is a Web document containing academic information about Stanford University. The initial *http* displays HTTP and the final *html* shows that the document is in the form of *hyper text mark-up language.* We will analyze the structure of the Web as a directed graph and then investigate models and algorithms using the graph representation of the Web.

12.3.1 The Web Graph

A Web page may reference another document (URL) and we can show this referencing by a directed line from the referencing page to the referenced page. We can therefore represent Web as a directed graph where nodes are the Web pages and the directed links are from the pages that reference to the pages that are referenced as shown in Figure 12.3.

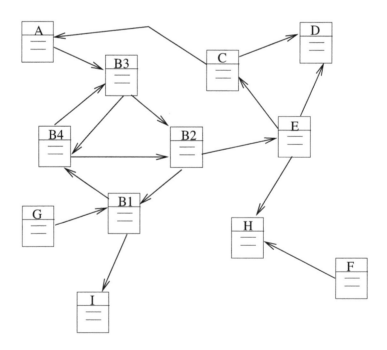

Figure 12.3: The Web as a directed graph

12.3.1.1 Properties

In the analysis of the Web graph, we need to consider the in-degrees and the out-degrees of the nodes as it is directed. Broder et al. experimented with 200M pages of Altavista in 1999 and showed the Web graph page in-degrees follow a power law distribution with $\gamma = 2.1$ [7]. The out-degree distribution of the nodes also showed power law distribution with an exponent approximately equal to 2.7. Donato et al. also found that the in-degree distribution of the Web graph follows the power law [20]. However, the out-degree distribution of the Web graph does not show a power law distribution, possibly due to lack of hyperlinks pointing to other documents in Web pages [36]. Web sites instead of Web pages were used as the nodes of the Web graph by Adamic et al. in [1]. A directed link from a site A to B showed that there is at least one Web page at site A referencing a document at site B. The average path length in this study was found as 3.1 and the average clustering coefficient was 0.11 which shows this alternative representation of the Web graph is still a small-world network. The search of substructures in the Web graph can be done at local or global levels. At local level, *bipartite cliques* where the nodes of a substructure is divided into two sets V_1 and V_2 such that each element in V_1 is connected to all elements of V_2 were found frequently in the Web graph in the study of Kumar et al. [31]. They discovered about a hundred thousand such communities. The properties of the Web graph can be summarized as follows [4]:

- *Dynamic nature*: New nodes are added and some nodes are deleted from the graph dynamically.

- *Power-law degree distributions*: The in-degree and the out-degree distributions follow power-law but with different characteristics.

- *Small-world property*: The average distance between the nodes is much smaller than the size of the graph.

- *Dense bipartite subgraphs*: The probability of finding distinct bipartite cliques or cores is larger than a random graph with the same number of nodes and edges.

The Bow-Tie Structure

A directed graph is *strongly connected* if there is a path between any pairs of nodes. A *strongly connected component* (SCC) of a directed graph $G(V,E)$ consists of the set of nodes $V' \in V$ such that there is a path from every node in V' to every other node in V' and also V' is not contained in any other SCC of G. The nodes $B1$, $B2$, $B3$ and $B4$ form a SCC in Figure 12.4. The bipartite cliques were assumed to be the *cores* of these communities. Broder et. al. provided a global map of the Web using strongly connected components as the basic building blocks [7] which has been and the outcome of their study was that the Web contains a giant strongly connected component. In practical terms, this would mean that major search engines reference major establishments such as institutions, agencies, companies which also reference

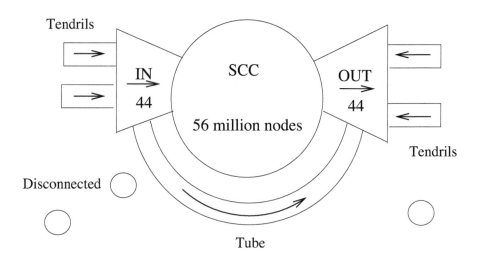

Figure 12.4: The bow-tie structure of the Web

the search engines and themselves. They also placed all of the other strongly connected components in relation to the giant component and found the following sets of nodes [7, 30]:

1. *IN*: These nodes can reach the giant component (*GC*) but they cannot be reached from it. Formally, given $G(V,E)$ as the Web graph, $\forall u \in IN$ and $v \in GC$, $\exists (u,v) \in E$ but $(v,u) \notin E$.

2. *OUT*: These nodes can be reached by the *GC* but they are not included in *GC*. Formally, $\forall u \in OUT$ and $v \in GC$, $\exists (v,u) \in E$ but $(u,v) \notin E$.

3. *Tendrils*: A tendril contains web pages that are connected to either *IN* or *OUT* but are not part of *IN*, *OUT* or *GC*.

4. *Disconnected*: These are the pages that cannot be accessed from any other groups. Starting from a disconnected page, we can never reach a page in *IN*, *GC* or *OUT*.

The *IN* set contains pages that have not been discovered by the pages in the giant component and *OUT* set has the pages that are referenced by the pages in the giant component but these pages do not reference pages in the giant component [7]. Relation between these components is shown in Figure 12.4. Broder et. al. discovered that the number of pages in *IN*, *OUT* and tendrils totaled 44 million whereas the giant component had about 44 million pages and approximately 17 million pages were disconnected. These figures meant that a significant proportion of total Web pages which are in *OUT* or in one of the tendrils cannot reach the giant component.

A realistic model of the Web graph should exhibit the properties we have outlined above. We will describe four important models based on these properties of Web graph, known as preferential attachment (evolving) model, copying model, growth-deletion model and multilayer model as described in [20].

12.3.1.2 Evolving Model

The first model for Web graph was proposed by Albert and Barabasi [3] with the *preferential attachment* (PA) rule which states that the new nodes joining the network prefer to get connected to nodes with higher degrees. They analyzed this model and showed that the graphs generated using this model have a power law distribution with $\gamma = 3$. Bollobas et al. provided an elaborate model based on the PA rule called the *Linearized Chord Diagram* (LCD) model [6] which is obtained by random pairings on fixed finite sets of integers. Aiello et al. also proposed four evolving graph models based on the PA rule where γ could be any value greater than 2 based on the input parameters [2]. The model of Cooper and Frieze used linear algebraic methods and is based on the PA rule [14]. The models proposed in [21, 22] are also based on the PA rule but they consider an *initial attractiveness* value and the probability of a new node to be connected to an existing node is proportional to the sum of its in-degree and this attractiveness value. In the model proposed by Pandurangan et al. [35], the evolving model is extended to include page rank of a node which is a value attributed to it based on the number of Web pages referencing it. A node is selected as the endpoint of an edge with probability $\alpha \in [0,1]$ proportional to its in-degree, and with probability $\beta \in [0,1]$ proportional to its page rank value and randomly with probability $1 - \alpha - \beta$ [35, 28]. The resulting graphs had the properties of in-degree and the page rank.

12.3.1.3 Copying Model

The copying model was proposed by Kumar et al. to improve the existing Web graph models [31]. In this model, when a new node u_t enters the Web graph at time t, a prototype node p is selected at random from the existing nodes. Assuming the Web graph at time t is $G_t(V_t, E_t)$ with constant d outgoing links at each node, endpoint of an outgoing link of node u_t is either copied with probability $\alpha \in [0,1]$ from the endpoint of an outgoing link of node p, or it is selected at random with probability $1 - \alpha$ from one of the existing nodes. The parameter α is called the *copying factor* and the aim of copying is to provide a model that describes the occurrences of many bipartite cliques in the Web graph.

Two versions of this model are the *linear growth copying* where a constant amount of nodes are added at each time step, and *the exponential growth copying* in which the Web graph is allowed to grow by a function of its size. This model provided in-degree and disjoint bipartite clique distributions following power law at $\gamma = 2.1$ when $\alpha = 0.8$ in the Web graph.

12.3.1.4 Growth-deletion Model

The *growth-deletion* models consider both the addition and the removals of nodes and edges during the generation of nodes of the Web graph [20] which in fact is a more realistic view of the evolving property of the Web. Bollobas et al. studied the effect of node and edge removals in their LCD model [6] to find the effects of random failures and random attacks after the graphs were generated. Chung proposed a growth-deletion model graph $G(p_1, p_2, p_3, p_4, m)$ where m is a positive integer, and $p_1, ..., p_4$ are the probabilities with the following conditions: $p_1 + p_2 + p_3 + p_4 = 1$; $p_3 < p_1$; $p_4 < p_2$ and the graph H is a fixed nonempty graph. Assuming a new node u_t at time t is to be added to the network, p_1 is the probability of adding node u and m edges incident with it to existing nodes in G_t using PA rule; p_2 is the probability of adding m edges to the existing nodes by the PA rule; p_3 shows the probability of deleting a node chosen uniformly at random from G_t, and p_4 is the probability of deleting m edges chosen uniformly at random from G_t. They showed that the degree distribution of the graph G_t generated by $G(p_1, p_2, p_3, p_4, m)$ follows a power law with probability 1 as $t \to \infty$ with the exponent as follows:

$$\gamma = 2 + \frac{p_1 + p_2}{p_1 + 2p_2 - p_3 - 2p_4} \tag{12.2}$$

In another growth-deletion model proposed by Cooper et al. [14], a dynamically evolving random graph was studied where nodes and edges are added using PA rule and nodes are deleted randomly. A new node u_t and m edges incident with it are added to existing nodes with probability $p_1 > 0$; with probability $p - p_1 \geq 0$, m random edges to existing nodes are connected with probability proportional to the degree of nodes. With probability $1 - p - p_0$, a random node is deleted and if nodes exist in the graph, and m random edges are deleted with probability p_0. They showed for large k, t values, the expected number of nodes with degree k is approximately $d_{k,t}$ where as $k \to \infty$, $d_k \approx Ck^{-1-\gamma}$, with constant $C > 0$ and γ defined as follows:

$$\gamma = \frac{2(p - p_0)}{3p - 1 - p_1 - p_0} \tag{12.3}$$

12.3.1.5 Multi-layer Model

Dill et al. showed that the Web graph revealed a fractal structure in various forms [15]. They considered the Web graph as the result of many similar and independent stochastic processes. *Cohesive collections* of pages at various scales exist, such as pages on a site or pages about a topic. These collections show similar structures to the Web and the central areas of these collections are called *thematically unified clusters* (TUCs) which provide a navigational backbone of the Web. Based on this concept, Laura et al. proposed the multi-layer model of the Web graph [34] with the following rules. Any new page that enters the Web graph is assigned a number of layers it will belong to and it can only get connected to nodes in these layers. A combination of evolving and copying models are used to assign links at each layer.

This model captured the power-law distribution of in-degrees for a range of variations of parameters.

Guido et al. provided a range of semi-external algorithms to discover the properties of the Web graph we have outlined above. Namely, they described algorithms to compute all SCCs and the largest SCC of the Web graph, random graph generators using the copying, evolving, multi-layer and page rank models [28].

12.3.1.6 Cyber Community Detection

There are many communities in the Web such as individuals with common interests and news groups. Other than these explicitly defined communities, there exist implicitly defined communities which are more difficult to identify. Discovery of these implicitly defined communities is needed as they provide information for the users interested in them. This information is not readily available in these groups as in the explicitly defined communities. For example, we may be interested to find the list of Chinese restaurants in Japan which may not be available as an explicit site. Detection of these communities will also give insights to the intellectual development of the Web.

Kumar et al. observed websites that are part of the same community do not usually reference each other, possibly due to the existences of competition between them and lack of a common point of view [31]. Another observation was that pages that are related are frequently referred together. Based on these observations, they proposed a model of the Web graph where the communities manifest themselves as bipartite graphs. A *core* is defined as a complete bipartite graph which is a subgraph of the bipartite graph representing a community in the Web. Assuming a community C_i is represented by a bipartite graph $G_i(V_1, V_2, E)$, they proposed to find $G_i^c \subset G_i$ to discover C_i and make use of G_i^c to discover G_i which can be used to discover C_i. A random bipartite graph $G(V_1, V_2, E)$ can be obtained by assigning m edges between V_1 and V_2 at random under any probability distribution. It can be shown that this random bipartite graph G contains a complete bipartite graph that is a subset of G. They also found that the in-degree of most of the Web pages is less than 410 and the probability that a page has in-degree k is $1/k^2$.

In the community detection algorithm they proposed, they considered the bipartite graph to be searched consisted of *fans* at one side and the *centers* to which the fans are connected at the other side. They discarded pages with in-degrees higher than a threshold k_1, as as these pages may be referenced for reasons such as popularity and may not actually indicate a specific community that is searched. They further discarded nodes that have an in-degree smaller than a threshold k_2 and an out-degree smaller than k_2. This pruning could result in more fans and centers being discarded as removing a fan with few outgoing links will decrease the in-degree of a center and conversely, removing a center with few incoming links will decrease the number of outgoing links of a fan. Further pruning is provided by the *inclusion-exclusion heuristic* which resulted in a total of unpruned 5 million pages to be searched in total. In the last step of the algorithm, these pages were explicitly enumerated to discover cores and the result was 75 K cores of sizes (3,3) [31]. The communities were then expanded using the hubs and authority scores of the nodes.

Other heuristic approaches for community detection include the method proposed by Flake et al. which use the maximum flow concept [24]. Gibson et al. used a sampling method called *shingling* which evaluates the similarity of the neighborhoods of nodes to detect large dense subgraphs of the Web graph [25].

12.3.2   Link Analysis

We will describe the link analysis of the Web as in [18]. When a keyword is entered using a search engine in the Web, we observe a list of a number of pages related to this keyword, some being very relevant and a number of them may not be so relevant to what we are searching. For example, when we enter the keyword "university" in a search engine, we will get a very large list output from any search engine. However, this list may contain similar elements such as the University of Cambridge and various other universities. One way of accessing the importance of an element in the output list is to count the number of pages that reference to it. In our university example, if we find University of Cambridge is pointed by many entries in the list, then we can conclude that it is more important than any other university we are searching. In order to account for this, we can simply count the votes received by the output pages by the pages that reference them, to find the scores showing their importance. We will describe formal procedures to asses the importance of pages based on this reasoning next.

12.3.2.1   Hubs and Authorities

The pages that cast votes are called *hubs* and some of these hubs that point to universities that are pointed by various other hubs can be considered important. In our example, some hubs listed in response to our query "university" will contain a list of universities, for example the list of the universities in the world and the Wikipedia university entry. In order to assign a score of importance to these hubs, we can find the set of pages they point to and calculate the sum of the votes the pages in the set received from all hubs. The pages that are pointed by the hubs which are the answers to our query are called *authorities*. Figure 12.5 depicts an example of the hubs and the authorities that may be obtained during a single query in the Web.

In order to assess the authority and the hub scores of a page p formally, Kleinberg proposed two rules as follows [30]:

■ *Authority Update Rule*: For each page p, its authority score $auth(p)$ is updated to yield the sum of the hub scores of all pages pointing to it.

■ *Hub Update Rule*: For each page p, its hub score $hub(p)$ is updated to be the sum of the authority scores of all pages that it points to.

We can implement these two rules for the example in Figure 12.5 to obtain the hub and authority scores shown next to the pages in the same figure. We can continue

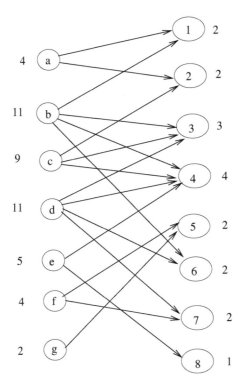

Figure 12.5: Hubs and authorities example.

with this process by implementing the authority rule first and then the hub update rule second for a number of times to refine the results obtained. The resulting scores may be large but we can normalize them by dividing each hub score by the square root of the sum of the squares of all hub scores, and dividing each authority score by square root of the sum of the squares of all authority scores. We will divide the hub score of a page by the sum of all hub scores and the authority score of a page by the sum of all authority scores for simplicity. Kleinberg proposed an algorithm called hypertext-induced topic selection (HITS) shown in Alg. 12.1, to implement the described procedure and showed that the authority and hub scores converge as the number of iterations go to infinity [18].

Table 12.1 displays the first two iterations of this algorithm in our example query graph and we can see that the scores start converging even after two steps. We can conclude that the authorities 3, 4 and 6 have much higher scores than the rest of the authorities and the hubs *b*, *c* and *d* have significantly higher scores than the other hubs in this sample. We could have easily detected visually that these authorities have received more votes than others and also the hubs *b*, *c* and *d* have voted for authorities that have received more votes. However, for large graphs obtained as a result of a single query, this would be very difficult.

Algorithm 12.1 *HITS_Alg*

1: **Input** : $P = \{p_1, ..., p_n\}$ ▷ set of n pages
2:      k steps
3: **Output** : authority and hub values for all pages
4: **for all** $p \in P$ **do**   ▷ initialize authority and hub values
5:      $hub_p \leftarrow 1; auth_p \leftarrow 1$
6: **end for**
7: **for** $j \leftarrow 1$ to k **do**   ▷ implement rules for k steps
8:      **for all** $p \in P$ **do**
9:        **apply** *Authority Update Rule* to p to get $auth_p$
10:      **end for**
11:      **for all** $p \in P$ **do**
12:        **apply** *Hub Update Rule* to p to get hub_p
13:      **end for**
14:      $auth\_sum \leftarrow \sum_{p \in P} auth_p$ ▷ find sums of values
15:      $hub\_sum \leftarrow \sum_{p \in P} hub_p$
16:      **for all** $p \in P$ **do**
17:        $auth_p \leftarrow auth_p / auth\_sum$ ▷ normalize values
18:        $hub_p \leftarrow hub_p / hub\_sum$
19:      **end for**
20: **end for**

Table 12.1: Hub and Authority Scores

	Hub	a	b	c	d	e	f	g	-	Total
	Auth.	1	2	3	4	5	6	7	8	
$k=1$	Auth.	2	2	3	4	2	2	2	1	18
	Hub	4	11	9	11	5	4	2	-	46
$k=1$ n	Auth.	0.11	0.11	0.17	0.22	0.11	0.11	0.11	0.06	
	Hub	0.09	0.24	0.20	0.24	0.11	0.05	0.04	-	
$k=2$	Auth.	0.33	0.29	0.68	0.79	0.05	0.48	0.29	0.04	2.95
	Hub	0.62	2.28	1.76	2.24	0.83	0.34	0.05	-	8.12
$k=2$ n	Auth.	0.11	0.10	0.23	0.27	0.02	0.16	0.10	0.01	1.00
	Hub	0.08	0.28	0.22	0.28	0.10	0.04	0.01	-	1.01

Analysis

We can form the adjacency matrix for the pages in the Web graph which contains a unity entry in the ith row and jth column if page i has a directed link to page j. The hub scores of n pages in the Web graph can be stored in a vector $H[n]$ where the ith entry corresponds to the hub score for page i. Similarly, the vector $U[n]$ stores the authority values for n pages. Given the directed graph $G(V, E)$ of n Web pages with an adjacency matrix A, we can formulate the hub update rule with the following matrix equation:

$$H[n] = A[n,n] \times U[n] \qquad or \qquad H = AU \qquad (12.4)$$

Writing this equation for our example in Figure 12.5 yields:

$$
\begin{bmatrix} 4 \\ 11 \\ 9 \\ 11 \\ 5 \\ 4 \\ 2 \end{bmatrix} = \begin{bmatrix} 1 & 1 & 0 & 0 & 0 & 0 & 0 & 0 \\ 1 & 0 & 1 & 1 & 0 & 1 & 0 & 0 \\ 0 & 1 & 1 & 1 & 0 & 0 & 0 & 0 \\ 0 & 0 & 1 & 1 & 0 & 1 & 1 & 0 \\ 0 & 0 & 0 & 1 & 0 & 0 & 0 & 1 \\ 0 & 0 & 0 & 0 & 1 & 0 & 1 & 0 \\ 0 & 0 & 0 & 0 & 1 & 0 & 0 & 0 \end{bmatrix} \times \begin{bmatrix} 2 \\ 2 \\ 3 \\ 4 \\ 2 \\ 2 \\ 2 \\ 1 \end{bmatrix} \quad (12.5)
$$

For the authority update rule, we need to sum the number of incoming edges to a page i, therefore, we need to use the transpose of matrix A as follows:

$$U = A^T H$$

In order to account for k steps:

$$U^1 = A^T H^0$$

$$H^1 = AU^1 = AA^T H^0$$
$$U^2 = A^T H^1 = A^T AA^T H^0$$
$$H^2 = AU^2 = AA^T AA^T H^0 = (AA^T)^2 H^0$$

It becomes clear that the general rule is:

$$U^k = (A^T A)^{k-1} A^T H^0$$
$$H^k = (AA^T)^k H^0$$

which means that the authority vector U and the hub vector H at kth step are obtained by taking powers of the products $A^T A$ and AA^T in the order of k [18] .

12.3.2.2 *Page Rank Algorithm*

We have analyzed a dynamically formed Web graph in response to a query. This bipartite graph contained two different type of pages as hubs and authorities. The Web graph in general is not bipartite and can be viewed as a directed graph where the bipartite structure may exist only locally. *Page rank* is an attribute of importance of a page in the Web graph based on the number of pages that reference it. It is basically a score for a page based on the votes it receives from other pages. This is a sensible metric for the importance of a page since the relative importance and popularity of a page increase by the number of pages referencing it. Page rank can be considered as a fluid that runs through the network accumulating at important nodes. The page rank algorithm to find importance of pages in a Web graph assigns ranks of the pages in the Web graph such that the total page rank value in the network remains constant. It initially assigns rank values of $1/n$ to each page in an n node network as shown in Alg. 12.2. The current rank value of a page is evenly distributed to its outgoing links

---

Algorithm 12.2 *Page_Rank_Alg*

1: **Input** : $P = \{p_1, ..., p_n\}$ ▷ set of *n* pages
2:       *k* steps
3: **Output** : page rank values $rank_p$, $p \in P$
4: $E_p(in) \leftarrow$ ingoing edges to page *p*
5: $E_p(out) \leftarrow$ outgoing edges from page *p*
6: **for all** $p \in P$ **do**   ▷ initialize page rank values
7:      $rank_p \leftarrow 1/n$
8: **end for**
9: **for** $r \leftarrow 1$ to *k* **do**   ▷ implement for *k* steps
10:      **for all** $p \in P$ **do**
11:         **for all** $e \in E_p(out)$ **do**
12:            $w_e \leftarrow rank_p/|E_p(out)|$
13:         **end for**
14:         $rank_p \leftarrow \sum_{e \in E_p(in)} w_e$ ▷ the sum of the weights of all links pointing to p_i
15:      **end for**
16: **end for**

---

and then, the new page rank values are calculated as the sum of the weights of the ingoing links of pages.

Execution of this algorithm for *k* steps results in more refined results for page rank values as in the authority and hubs algorithm and the page rank values converge as $k \rightarrow \infty$. The implementation of this algorithm for two iterations in the Web graph of Figure 12.6 which shows the initial weights assigned to edges, yields the page rank values shown in Table 12.2.

We see that pages 4 and 5 have significantly higher rank values than other pages. This is due to page 4 being referenced by the only outgoing edge from page 5, and page 5 being referenced by the only outgoing edge from page 1. There is a problem with the page rank algorithm which is more evident when there are a number of pages that can be reached from all nodes of the Web graph but they do not have many outgoing links. These nodes tend to accumulate page rank scores as the number of iterations of the algorithm grows. In order to remedy this situation, all page rank values can be scaled down by a factor of *d* called the *damping factor*, so that the nodes are assigned $(1-d)/n$ values initially [30]. In practice, the scaled page rank algorithm is used with *d* having a value between 0.8 and 0.9 .

Spectral Analysis

In order to analyze the page rank algorithm, we define a matrix *W* for *n* pages with an entry w_{ij} which is the share of page *i* rank value that page *j* should get in the update procedure. This would mean w_{ij} is equal to zero if page *i* does not point to page *j* and is equal to r_i/m_i if it points to *j*, r_i being the current page rank value of page *i* and m_i is the number of the outgoing links of page *i*. Initially, the $1/n$ value of a page

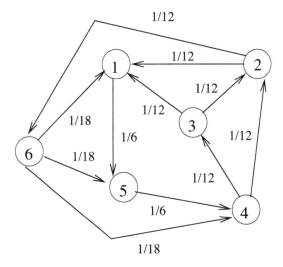

Figure 12.6: An example Web graph

Table 12.2: Page Rank Values

		1	2	3	4	5	6
n_{out}		1	3	2	2	1	3
k= 1	weight/edge	1/6	1/12	1/12	1/12	1/6	1/18
rank		2/9	1/6	1/12	2/9	2/9	1/12
k= 2	weight/edge	2/9	1/12	1/24	1/9	2/9	1/36
rank		11/72	11/72	1/9	1/4	1/4	1/12

is divided equally between its outgoing edges. For the example Web graph of Figure
12.6 allows us to form the matrix W as follows:

$$
\begin{bmatrix}
0 & 0 & 0 & 0 & 1/6 & 0 \\
1/12 & 0 & 0 & 0 & 0 & 1/12 \\
1/12 & 1/12 & 0 & 0 & 0 & 0 \\
0 & 1/12 & 1/12 & 0 & 0 & 0 \\
0 & 0 & 0 & 1/6 & 0 & 0 \\
1/18 & 0 & 0 & 1/18 & 1/18 & 0
\end{bmatrix}
\tag{12.6}
$$

In this notation, the jth column of the matrix W corresponds to the page rank
values that page j receives. Therefore, adding all the elements of a column j of W
yields the page rank value of page j. The page rank update rule can then be specified
as follows:

$$
R^k \leftarrow W^T[1]
\tag{12.7}
$$

We can write this equation for the example graph of Figure 12.6 for the first iteration to yield:

$$
\begin{bmatrix}
2/9 \\
1/6 \\
1/12 \\
2/9 \\
2/9 \\
1/12
\end{bmatrix}
=
\begin{bmatrix}
0 & 1/12 & 1/12 & 0 & 0 & 1/18 \\
0 & 0 & 1/12 & 1/12 & 0 & 0 \\
0 & 0 & 0 & 1/12 & 0 & 0 \\
0 & 0 & 0 & 0 & 1/6 & 1/18 \\
1/6 & 0 & 0 & 0 & 0 & 1/18 \\
0 & 1/12 & 0 & 0 & 0 & 0
\end{bmatrix}
\times
\begin{bmatrix}
1 \\
1 \\
1 \\
1 \\
1 \\
1
\end{bmatrix}
\tag{12.8}
$$

Alternatively, we can define another matrix $M[n,n]$ which has elements $m_{ij} = 1/l_i$ where l_i is the number of the outgoing links from page i. In this case, the basic page update rule can be stated as:

$$
R^{k+1} \leftarrow M^T R^k \tag{12.9}
$$

where R at the right hand side is the vector of old page rank values to compute the new rank values in the R vector in the left. The scaled update rule is similar:

$$
R^{k+1} \leftarrow \overline{M}^T R^k \tag{12.10}
$$

where an element of $\overline{M}$ is $\overline{m}_{ij} = m_{ij} + (1-d)/n$. For the kth iteration, this equation can be written as:

$$
R^k = (\overline{M}^T)^k R^0 \tag{12.11}
$$

where R^0 is the initial page rank vector of $1/n$ values. We do not need to normalize these values as the sum of the page rank values is constant. Based on Perron's theorem which states that for any matrix A with all positive elements, there is a positive eigenvector y corresponding to the largest eigenvalue [33]. If the largest eigenvalue is 1, then starting from any initial positive vector x, $A^k x$ converges to a vector in the direction of y as $k \to \infty$. This would mean that repeated application of the scaled page rank update rule results in a converged rank vector R^E which satisfies the equation $R^E = (\overline{M}^T)R^E$ with R^E being an eigenvector of $\overline{M}^T$ with eigenvalue 1. In other words, starting from any initial configuration, we will get a converged page rank vector R^E [18].

We have described two different methods of assessing the importance of pages in the Web as HITS and the page rank algorithm. A comparison between these two algorithms are as follows:

■ HITS depends on the query but page rank is independent of the query.

■ Page rank is applied to all of the pages in the Web whereas HITS provides values in the local neighborhood of the query results.

■ Page rank has a problem with pages that do not have many outgoing links as we saw, but this can be corrected by the use of scaling down the values.

■ Modifications to links may change the obtained page values significantly in both algorithms.

12.4   Chapter Notes

In this chapter, we have described the Internet structure, the protocols used, and then analyzed its properties based on experiments performed in various projects. These research studies used the routing table information stored at the nodes or the software probes that monitor the status of the network. The results from these tests showed that the Internet is scale-free and follows the small-world model.

We then investigated the structure of the Web which is commonly modeled as a directed graph of Web pages. We have also described more sophisticated models of the Web graph as the evolving, copying, growth-deletion and multi-layer models. The experiments performed on the Web show that it is also a scale-free network with small-world characteristics like the Internet. An important result of the experimental evaluation of the Web showed that it has a bow-tie structure. The hubs and authorities algorithm (HITS) is a simple procedure that assigns importance values to pages that reference a page and the pages that are referenced. The page rank algorithm described is widely used in the Web to find the rank of a Web page. The main differences between these two algorithms is that HITS is based on a query and is used to compute authorities and hubs to classify and rank the retrieved data from a query whereas page rank algorithm considers the entire Web.

The Internet and the Web are both large and complicated complex networks. A further complexity is encountered when the dynamic and the evolving natures of these networks are considered. Testing and evaluation of the results of the experiments for these two networks are difficult tasks as testing of the whole network is impossible in general. Any test procedure should therefore start with a representative sample of the network which means sampling and also modeling of these large networks are two fundamental tasks before obtaining any meaningful results. Modeling a dynamic network is different than modeling a static one as the nature and structure of a dynamic network changes significantly with time. Obtaining test data either by examining local node data or monitoring network activity is another issue to be handled. Once meaningful results are obtained, we need to analyze these and use statistical methods to explain the behavior of these two large networks before reaching any reliable results. We can therefore state that sampling, modeling, obtaining reliable data and then analysis of this data are the key problems to be addressed in the study of these networks in current and future studies.

Exercises

1. Discuss the properties of the Internet as a complex network.

2. Compare the two fundamental Internet protocols TCP and UDP in terms of their applications.

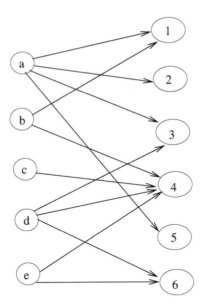

Figure 12.7: Example graph for Ex. 3

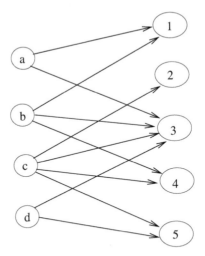

Figure 12.8: Example graph for Ex. 4

3. Work out the authority and hub values of the bipartite Web graph of Figure 12.7 for three iterations using the hub and authority algorithm. Calculate the normalized values at each step and discuss the results.

4. Find the hubs and authority values of the pages in the graph of Figure 12.8,

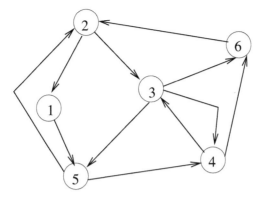

Figure 12.9: Example graph for Ex. 5

this time using the matrix equations. Work out the normalized values for four iterations and plot these values.

5. Find the page rank values for the example Web graph of Figure 12.9 using the page rank algorithm and tabulate your results.

References

[1] L. Adamic. The Small World Web. In Abiteboul S. and Vercoustre A.-M., Eds., ECDL, Volume 1696 of Lecture Notes on Computer Science, pages 443-452, Berlin, Sept. 1999. Springer-Verlag.

[2] W. Aiello, F. Chung, and L. Lu. Random evolution in massive graphs. *Handbook on Massive Data Sets*, (James Abello et al., Eds.), Kluwer Academic Publishers, 97-122, 2002.

[3] R. Albert and A. Barabasi. Emergence of scaling in random networks. *Science*, 286:509-512, 1999.

[4] A Bonato. A survey of models of the Web graph. In *Combinatorial and Algorithmic Aspects of Networking*, Springer Lecture Notes in Computer Science, Volume 3405, pages 159-172, 2005.

[5] R. Bellman. On a routing problem. *Quarterly of Applied Mathematics*, 16:87-90, 1958.

[6] B. Bollobas, O. Riordan, J. Spencer and G. Tusnady. The degree sequence of a scale-free random graph process. *Random Structures Algorithms*, 18:279-290, 2001.

[7] A.Z. Broder, S.R. Kumar, F. Maghoul, P. Raghavan, S. Rajagopalan, R. Stata, A. Tomkins, and J.L. Wiener. Graph structure in the Web. In *Proceedings of the 9th International World Wide Web Conference on Computer Networks*, 33(1-6):309-320, 2000.

[8] G. Caldarelli and A. Vespignani. *Large Scale Structure and Dynamics of Complex Networks: From Information Technology to Finance and Natural Science.* (Complex Systems and Interdisciplinary Science), World Scientific Publishing Company, Chapter 8, ISBN-13: 978-9812706645, 2007.

[9] Router-Level Topology Measurements: Cooperative Association for Internet Data Analysis. http://www.caida.org/tools/measurement/skitter/routertopology/.

[10] C. Cooper, A. Frieze, and J. Vera. Random deletion in a scale-free random graph process. *Internet Mathematics*, 1:4, 463-483, 2004.

[11] C. Cooper and A. Frieze, On a general model of web graphs. *Random Structures and Algorithms* 22(3):311-335, 2003.

[12] K.M. Chandy and J. Misra. Distributed computation on graphs: Shortest path algorithms. *Commun. ACM*, (25)11:833-838, 1982.

[13] F. Chung and L. Lu. The average distances in random graphs with given expected degrees. *Internet Mathematics* 1:91-114, 2003.

[14] C. Cooper , A. Frieze, and J. Vera. Random deletion in a scale-free random graph process. *Internet Mathematics*, 1(4):463-483, 2004.

[15] S. Dill, R. Kumar, K. McCurley, S. Rajagopalan, D. Sivakumar, and A. Tomkins. Self-similarity in the Web. In *Proceedings of the 27th International Conference on Very Large Databases*, pages 69-78. San Francisco: Morgan Kaufmann, 2001.

[16] Y. Shavitt and E. Shir. Dimes: Let the Internet measure itself. *Computer Communication Review*. 35(5):71-74, 2005.

[17] Distributed Internet Measurements and Simulations. http://www.netdimes.org.

[18] D. Easley and J. Kleinberg. *Networks, Crowds, and Markets: Reasoning about a Highly Connected World.* Cambridge University Press, Chapter 14, 2010.

[19] G.W. Flake, S. Lawrence, C.L. Giles, and F. Coetzee. Self-organization of the Web and identification of communities. *IEEE Computer*, 35(3):66-71, 2002.

[20] D. Donato, L. Laura , S. Leonardi, and S. Millozzi. The Web as a graph: how far we are. *ACM Transactions on Internet Technology*, 7(1), 2007.

[21] S.N. Dorogovtsev, J.F.F. Mendes, and A.N. Samukhin. Structure of growing networks with preferential linking. *Physical Review Letters*, 85:4633-4636, 2000.

[22] E. Drinea, M. Enachescu, and M. Mitzenmacher. Variations on random graph models for the Web. Technical report, Department of Computer Science, Harvard University, 2001.

[23] M. Faloutsos, P. Faloutsos, and C. Faloutsos. On power-law relationships of the Internet topology. *Computer Communication Review*, 29(4):251-262, 1999.

[24] G.W. Flake, S. Lawrence, C.L. Giles, and F. Coetzee. Self-organization of the Web and identification of communities. *IEEE Computer*, 35(3):66-71, 2002.

[25] D. Gibson, R. Kumar, and A. Tomkins. Discovering large dense subgraphs in massive graphs. In *Proceedings of VLDB '05*, 721-732, 2005.

[26] http://www.ams.org/samplings/feature-column/fcarc-pagerank

[27] R. Govindan and H. Tangmunarunkit. Heuristics for Internet map discovery. In *Proceedings of IEEE INFOCOM'00*, pages 1371-1380, 2000.

[28] C. Guido and A. Vespignani. *Large Scale Structure and Dynamics of Complex Networks: From Information Technology to Finance and Natural Science*, Chapter 7, World Scientific Publishing Company, 2007.

[29] University of Oregon Route Views Project, http://www.routeviews.org/.

[30] J. Kleinberg. Authoritative sources in a hyperlinked environment. *Journal of the ACM*, 46(5):604-632, 1999.

[31] R. Kumar, P. Raghavan, S. Rajagopalan and A. Tomkins, Trawling the Web for emerging cyber-communities. In *Proceedings of the 8th WWW Conference*, Elsevier North-Holland, Inc., pages 1481-1493, 1999.

[32] J. Kurose and K. Ross. *Computer Networking: A Top Down Approach*, Pearson; 6th edition, ISBN-10: 0132856204, ISBN-13: 978-0132856201, 2012.

[33] A.N. Langville and C. D. Meyer. *Googles PageRank and Beyond: The Science of Search Engine Rankings*. Princeton University Press, 2006.

[34] L. Laura, S. Leonardi, G. Caldarelli and P.D.L. Rios, A multi-layer model for the webgraph. In *On-line Proceedings of the 2nd International Workshop on Web Dynamics*, 2002.

[35] G. Pandurangan, P. Raghavan and E. Upfal. Using page rank to characterize Web structure. Preliminary version in *Proceeding of 8th Annual International Conference on Combinatorics and Computing*, Lecture Notes in Computer Science, 2387, Springer-Verlag, pages 330-339, 2002.

[36] M. van Steen. *Graph Theory and Complex Networks: An Introduction*. ISBN-13: 978-9081540612, 2010.

Chapter 13

Ad hoc Wireless Networks

13.1 Introduction

A wireless network consists of computing nodes which communicate using wireless communication channels. Wireless networks can be formed as *infrastructured* or *ad hoc*. A static wired backbone usually consisting of host computers provides the communication in infra-structured networks whereas nodes of an ad hoc wireless network communicate using multi-hop packet transfers. Two important types of ad hoc wireless networks are the mobile ad hoc networks (MANETs) and wireless sensor networks (WSNs). There is no fixed infrastructure in a MANET as the nodes in such a network change their positions dynamically. Nodes in a MANET communicate with their current immediate neighbors to transfer messages to destinations. Figure 13.1 displays a MANET with host computers $A, ...H$ which communicate using wireless links. It can be seen that mobile hosts F and G have a more central function than others as most of the message transfers between any nodes should pass through them.

Routing is the process of determining the lowest cost paths between source nodes and destination nodes in a computer network. Since the topology of a MANET changes frequently, finding efficient routes dynamically is one of the main challenges in these networks. Another challenge is the provision of access control to the shared medium as concurrent accesses will result in collisions of packets and should be avoided.

A WSN consists of hundreds and thousands of sensor nodes each equipped with a battery and antenna to communicate with neighbors. These sensor nodes gather data from the surrounding area and send this data to a special node called the *sink*, which has better processing capabilities and performs further fine processing of the data and may transfer it to a remote host computer for analysis. Figure 13.2 shows the general structure of a sensor network where nodes send their data to the sink

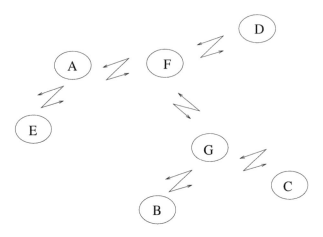

Figure 13.1: A MANET example

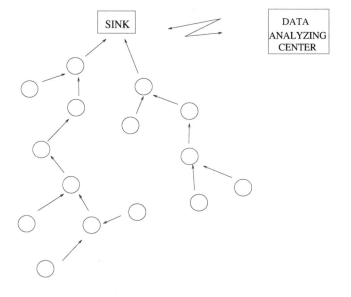

Figure 13.2: A WSN example

using a spanning tree structure and the sink transfers the data to a data analyzing center after initial processing. Different from a MANET, data transfer is oriented toward the sink and routing of data packets to the sink in this manner is an important issue in a WSN. Furthermore, limited lifetime of the on-board batteries necessitates employment of energy efficient algorithms with low message complexities for routing and other tasks as message transfers are the main source of energy dissipation in WSNs as in MANETs. In this chapter, we will review the structures of MANETS and WSNs and clustering in these networks, and investigate mobile social networks that have attracted significant research efforts recently.

13.2   Clustering Algorithms

Clustering in MANETs and WSNs has different goals and is performed differently from clustering in other types of complex networks. Firstly, the clustering is performed by the individual nodes of the network and each node is aware of its cluster and usually the members of the cluster at the end of the process. We will call this type of clustering *distributed clustering* and describe sample distributed clustering algorithms for MANETs and WSNs in this section.

MANETs require low maintenance clustering algorithms as they need to be executed frequently due to the dynamicity of the nodes in such a network. A clusterhead (CH) of a cluster is a special node that manages membership in a cluster and can act on behalf of all nodes for functions such as routing. Selection of the optimum number of CHs is NP-hard, therefore heuristics using node degree, node identifier and mobility are usually employed to choose CHs. For load balancing and fault tolerance, CHs may be rotated among the members of the cluster. A *gateway* node is a node that connects at least two clusters, hence, three types of nodes in a clustered network are the CHs, gateway nodes and the ordinary nodes. Clustering in ad hoc wireless networks is commonly done in two steps: the cluster formation and the cluster maintenance. In a MANET, a new CH has to be elected dynamically for a group of nodes as they change their positions.

13.2.1   Lowest-ID Algorithm

The first algorithm we will investigate to form clusters around a CH in a MANET is called *lowest identifier algorithm* and is due to Gerla and Tsai [10]. In this algorithm, each node periodically broadcasts the nodes it can hear (detect) including itself, and the following rules are then applied:

1. A node decides to be a CH if it does not hear a node with a higher identifier than itself.

2. The lowest identifier neighbor that a node hears is marked by the node as its CH, unless that node voluntarily gives up its position as a CH.

3. A node that hears two or more CHs becomes a *gateway* that joins two clusters.

Alg. 13.1 shows a possible implementation of *Lowest_ID* algorithm [2]. We assume each node is aware of the identifiers of its neighbors which can be achieved by periodic sending of messages such as *"I am alive"* at MAC level. Each node periodically checks the identifiers of its active neighbors and if it finds it has the lowest identifier among all neighbors, it broadcasts that it is the CH and if it discovers it has two broadcasting neighbors, it assigns itself as a gateway and any other node becomes an ordinary node.

Algorithm 13.1 *Lowest_ID*

1: **set of int** *neighs*, *my_cheads* ← {Ø}
2: **states** *chead*, *gateway*, *ordinary*
3: **message types** *me_chead*, *me_ordinary*
4: **boolean** *has_chead* ← *false*
5: **loop**   ▷ do periodically
6:     **if** *my_id* = min{*neighs*} **then**
7:         **broadcast** *me_chead*
8:         *state* ← *chead*
9:     **else**
10:         **broadcast** *me_ordinary*   ▷ needed to check all neighbors
11:         *state* ← *ordinary*
12:     **end if**
13:     **while** *received* ≠ *neighs* **do**   ▷ receive neighbor info
14:         **receive** *msg(j)*
15:         **if** *msg(j).type* = *me_chead* **then**
16:             *my_cheads* ← *my_cheads* ∪ {*j*}
17:             **if** *has_chead* = *false* **then** *has_chead* ← *true*   ▷ first CH
18:             **else** *state* ← *gateway*   ▷ node is a gateway
19:             **end if**
20:         **end if**
21:         *received* ← *received* ∪ {*j*}
22:     **end while**
23: **end loop**

An example MANET that is divided into three clusters in Figure 13.3 with nodes as $C_1 = \{2,5,6,8,12\}$, $C_2 = \{1,3,7,8,11\}$, $C_3 = \{4,6,9,10,13\}$ with CH nodes 2, 1 and 4 for clusters C_1, C_2 and C_3 respectively. Nodes 8 and 6 are the gateway nodes between clusters C_1 and C_2; and C_1 and C_3 respectively. Total number of messages communicated in Alg. 13.1 is n as each node will broadcast one message (*me_chead* or *me_ordinary*) to inform whether it is a CH or an ordinary node to its neighbors. The *Lowest_ID* algorithm may result in unstable CH selection since a low identifier node entering an already formed cluster may have all of the structure of the cluster to be re-organized whereas it could simply have joined the cluster.

Symmetries can be broken by the degrees of nodes rather than the identifers to elect the CHs. In this case, each node broadcasts its current degree periodically and a node that finds it has the highest degree among all of its neighbors declares itself as the CH in its neighborhood as proposed in [10]. The ties can be resolved using the identifiers where a lower identifier node with the same degree wins. Gerla and Tsai

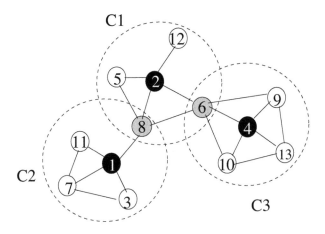

Figure 13.3: a) Lowest identifier algorithm. Clusterheads are shown in bold, gateway nodes are shown in gray. Transmission ranges of CHs are shown by dashed circles.

proposed a second algorithm to solve the unstability problem described [11]. This algorithm is also based on lower identifiers of nodes and produces non-overlapping clusters. A k-hop cluster is a set of nodes within at most k-hop distance from their CH. Nocetti et al. [20] provided an algorithm to find k-hop clusters based on the second algorithm of Gerla and Tsai.

13.2.2 Dominating Set-based Clustering

Dominating sets can be used to construct clusters in any network and a node that is a member of the dominating set becomes a CH and the nodes that it dominates are the members of its cluster as we have seen in Chapter 8. The same idea can be implemented in MANETs and WSNs to form clusters in these networks. Moreover, if the dominating set formed is connected (CDS), we can use this CDS as a virtual backbone for routing purpose where any node that wants to send a packet to a destination simply sends the packet to its CH which then initiates the propagation of the packet in the virtual backbone of CDS nodes.

CDS algorithms for ad hoc networks can be maximal independent set (MIS)-based or non-MIS-based. In MIS-based algorithms such as in [2], a two-step algorithm first constructs an MIS and its members are connected to form a CDS in the second step. A different approach is taken in the algorithm of Wu et. al. as described next.

Pruning-based Algorithm

Wu and Li proposed a distributed algorithm to find a connected dominating set of a network which works in two steps [22]. Each node initially enters the CDS if it has two unconnected neighbors. This initial step may result in many nodes entering the CDS than necessary. The algorithm then applies two pruning rules to exclude some of these nodes from the CDS. In the first rule, a node checks whether a neighbor CDS node with a higher identifier covers its entire closed neighbor set and if there is such a node, it exits the CDS. The second rule forces any node which has an open neighborhood set covered by the union of the open neighborhoods of its two CDS neighbors which both have higher identifers than itself to exit the CDS. Assuming a node u in the CDS has color black and all others are white, the rules for u can be specified formally as follows:

- **If** $\exists v \in N[u] | (color(u) = color(v) = black) \wedge (N[u] \subseteq N[v]) \wedge (id(u) < id(v))$ **then** $color(u) \leftarrow white$.

- **If** $\exists v, w \in N(u) | (color(u) = color(v) = color(w) = black) \wedge (N(u) \subseteq (N(v) \cup N(w))) \wedge (id(u) = min\{id(u), id(v), id(w)\})$ **then** $color(u) \leftarrow white$.

Alg. 13.2 shows a possible distributed implementation of this algorithm for a node u assuming each node exchanges messages with its neighbors periodically about the neighbors of their neighbors. A received message is from node v and *colors* set holds the identifers of the neighbors with their respective colors.

Figure 13.4 shows the applications of the rules of this algorithm in two sample networks. In (a), nodes 1, 3, 4, 5 and 7 mark themselves black as they all have two unconnected neighbors. After exchanging a round of messages with neighbors, nodes 1 and 3 give up being in the CDS as their closed neighborhoods {1,4,5,7} and {3,5,6,7} are covered by the larger identifer nodes 5 and 7 which are black, respectively. In (b), nodes 1, 2, 3, 4 and 5 are in the CDS initially as they all have two unconnected neighbors, and applying Rule 2 results in nodes 1 and 2 to be excluded from the CDS as their open neighborhoods are covered by open neighborhoods of their two neighbors. For node 1, its neighbors 3, 4, 5 and 7 are covered by neighbors 3 and 4 which have higher identifiers than itself and are black; and for node 2, its neighbors 4, 5 and 6 are covered by neighbors 4 and 5 which are black and have higher identifiers. This algorithm has a message complexity of $O(m)$ as each edge is traversed twice, once in each direction after which rules are applied to form the CDS.

Cokuslu and Erciyes [7] provided an algorithm to improve the *Wu_MCDS* algorithm which considers the degrees of the nodes while pruning. In the first phase of the algorithm, the nodes which have isolated neighbors with no neighbors are marked black permanently. Other nodes with two unconnected neighbors are colored gray as in Wu's algorithm in this phase. The degrees of nodes along with their neighborhoods are compared in the second phase, while marking them black. The idea of this algorithm is that the nodes with higher degrees should have a better chance of being in CDS, resulting in smaller size CDS. They showed that this algorithm performs better than Wu's algorithm experimentally [7].

Algorithm 13.2 *Wu_CDS*

1: **int** *mycolor* ← *white*
2: **set of int** *neighs* ← *N(u)*, *received* ← ∅, *colors* ← ∅
3: **message types** *color*
4: **loop**
5:     **if** ∃*v*, *w* ∈ *N(u)*|(*v*, *w*) ∉ *E* **then** ▷ if two unconnected neighbors , enter CDS
6:         *my_color* ← *black*
7:     **end if**
8:     **send** *color(my_color)* to *N(u)*
9:     **while** *received* ≠ *N(u)* **do**
10:        **receive** *color(v_col)*
11:        *colors* ← *colors* ∪ {*v*, *v_col*}
12:        *received* ← *received* ∪ {*v*}
13:     **end while**
14:     **if** *my_color* = *black* **then**
15:         **if** (∃*v* ∈ *colors*|*color$_v$* = *black*) ∧ (*N(v)* ⊆ *N(u)*) ∧(*u* < *v*) **then** ▷ Rule1
16:             *my_color* ← *white*
17:         **end if**
18:         **if** (∃*v*, *w* ∈ *N(u)*|*color(v)* = *color(w)* = *black*) ∧ (*N(u)* ⊆ (*N(v)*) ∪ *N(w)*)) ∧ *id(u)* = *min(id(v), id(u), id(w))* **then**   ▷ Rule 2
19:             *my_color* ← *white*
20:         **end if**
21:     **end if**
22: **end loop**

13.2.3 Spanning Tree-based Clustering

An asynchronous single initiator spanning tree based clustering algorithm for WSNs was proposed by Erciyes et al. [9]. The sink node of the WSN is the initiator and a spanning tree rooted at the sink and clusters are built simultaneously during the

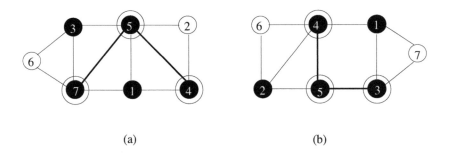

(a)   (b)

Figure 13.4: Wu's algorithm pruning rules. a) Rule 1. b) Rule 2. The final CDS nodes are shown in double circles

execution of this algorithm. Informally, the sink node starts the algorithm by sending a *probe*(*clust_id*, *hops*) message to its neighbors which then increment the hop count and broadcast *probe* message to their neighbors. Any node that receives this message for the first time assigns the sender as its parent and broadcasts it to its neighbors by incrementing the hop count. The implementation so far is identical to single initiator asynchronous building of a spanning tree in a graph. However, by testing of the hop count against a depth parameter *d*, clusters can be formed at the same time. Any node receiving the probe message for the first time also checks whether the *hops* parameter is equal to 0, and if so, the receiving node is the CH of the new cluster; if it is equal to *d* it is the *leaf* node of a cluster and any value between 0 and *d* results in an *intermediate* node. A leaf node places a 0 value for the *hops* value before sending the *probe* message to its neighbors. Alg. 13.3 shows the operation of the algorithm for node *u* where messages are received from node *v* and Figure 13.5 displays the clusters constructed in a sensor network using this algorithm with $d = 2$.

Algorithm 13.3 *ST_Clust*

1: **int** *parent* ← ⊥; *clust_id*, *cid*
2: **set of int** *childs* ← {∅} , *others* ← {∅}
3: **message types** *probe*, *ack*, *reject*
4: **states** *chead*, *intermed*, *leaf*
5: **if** *u* = *sink* **then**   ▷ root initiates tree construction
6:     **send** *probe*(*u*, 0) to *N*(*u*)
7:     *parent* ← *u*; *clust_id* ← *u*
8: **end if**
9: **while** (*childs* ∪ *others*) ≠ (*N*(*u*) \ {*parent*}) **do**   ▷ receive all neighbor msgs
10:     **receive** *msg*(*v*)
11:     **case** *msg*(*v*).*type* **of**
12:         *probe*(*clust_id*, *hops*)   : **if** *parent* = ⊥ **then**   ▷ *probe* first received
13:                                       *parent* ← *v*; **send** *ack* to *v*
14:                                       **if** *hops* = 0 **then**   ▷ i am chead
15:   *state* ← *chead*; *clust_id* ← *u*
16:                                       **else if** *n_hops* = *d* **then**
17:   *state* ← *leaf*; *clust_id* ← *cid*
18:                                       **else** *state* ← *intermed*; *clust_id* ← *cid*
19:                                       *hops* ← (*hops* + 1) MOD *d*
20:                                       **send** *probe*(*cid*, *hops*) to *N*(*u*) \ {*v*}
21:                                   **else send** *reject* to *v* ▷ *probe* received before
22:         *ack*   :   *childs* ← *childs* ∪ {*v*}   ▷ *v* is a child
23:         *reject*   :   *others* ← *others* ∪ {*v*}   ▷ *v* is unrelated
24: **end while**

The time required for the algorithm is the diameter *d* of the network. As the final spanning tree will have *n* nodes and *n* − 1 edges and each edge will have been

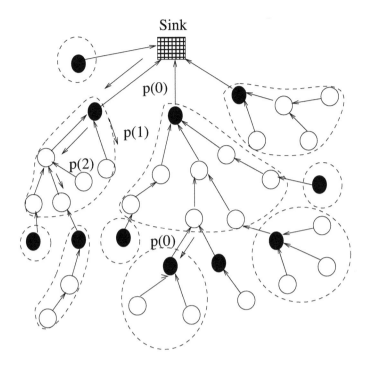

Figure 13.5: *ST_Clust* **execution in a sensor network**

traversed twice by *probe* and *ack* or *reject* messages, the total number of messages will be $O(m)$. Banerjee and Khuller [3] also proposed a protocol based on a spanning tree by grouping branches of a spanning tree into clusters of an approximate target size.

13.3 Mobile Social Networks

A *mobile social network* (MSN) is a mobile communication system where the users are involved in social interactions. Mobile users of these networks access, share and distribute information. Mobile social networks have a technological network component and a social network component. In terms of accessing and sharing data, these networks can be broadly classified as central and decentralized MSNs [15]. The users of central MSNs use Web-based services and communicate with the applications and each other using the Internet and the mobile devices. In the decentralized MSNs, a group of users share their data and communicate using a central server over a Wi-Fi connection.

MSNs provide dynamic exchange of information among people which can be analyzed by using the methods we have seen in Chapter 4. For example, based on the data collected, we can compute various centrality measures of an MSN. These networks are typically scale-free which means they have few large degree and many small degree components. They also exhibit small-world property where the number of communication links between any two users is relatively small compared with the size of the network.

Important applications of MSNs are the friendship networks which are social networks for people to get acquainted with each other and share information with each other. Many services for these MSNs are provided for mobile telephone users. Another application of MSNs is for healthcare services. In these MSNs, services provided to mobile patients and doctors enable continuous monitoring of patients and help can be provided when needed. An MSN can provide a location-based service where the location of a mobile user is determined and various facilities around the user such as friends, restaurants are informed to the user. We will now investigate technical aspects of MSNs and analyze them from a social network point of view.

13.3.1   Architecture

The main components of an MSN are the mobile users and devices, network infrastructure and the service and content providers. The network infrastructure provides the communication and data transfer between the users or devices and the servers. In centralized infrastructure, typically a Wi-Fi access point or a base station of a cellular network provides and maintains the connection to the server over the Internet as shown in Figure 13.6. The mobile user or device communicates with the central server using the client/server paradigm. Most of the on-line social networks use the centralized architecture.

In decentralized infrastructure, there is no notion of a central server and the mobile users communicate using Wi-Fi, cellular or Bluetooth interfaces. The decentralized infrastructure is sometimes called the *opportunistic network* where link performance is highly variable with possibly long delays [15]. Communication and data transfer among users is performed using multi-hop communication and access points. The hybrid architecture has both the centralized and distributed components as shown in Figure 13.7. There is a central server that provides the content but the users can also communicate among each other without using the server.

Cellular and Wi-Fi networks are the most common types of interface technologies for centralized MSN architectures. In opportunistic networks that use a distributed architecture, connectivity may be lost and routing of the packets is a fundamental problem in these networks. The *store-carry-forward routing* in distributed architectures where a network node stores data for a while before forwarding it to other nodes provides the data transfer in these networks.

13.3.2   Community Detection

Many social networks contain a community structure as we have seen in Chapter 11. These groups of densely interacting persons may be sharing common interests such as discussion topics, photographs or hobbies. Detecting these communities is

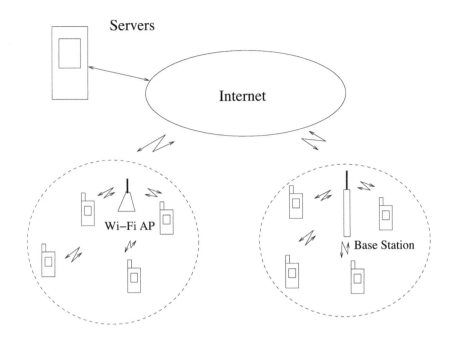

Figure 13.6: A centralized MSN

useful in various respects, for example, a community-aware routing protocol in an MSN may use packet transfers more efficiently by directing most of the messages to communities.

We can use any of the static community detection algorithms such as [16, 17] to discover these communities in MSNs by taking continuous snapshots of the network in time intervals whose durations are dependent on the characteristics of the network examined. Instead of using this time-consuming approach, we may update the network structure from its previous topology as outlined in [18]. Given an undirected unweighted graph $G(V,E)$ which represents the MSN, let $C = \{C_1, C_2, ..., C_k\}$ represent the possibly overlapping set of communities in G. We will further represent the initial network by $G_0(V_0, E_0)$ and the network at time t by $G_t(V_t, E_t)$. The network at time $t+1$ can therefore be represented by the graph $G_{t+1} = G_t \cup \Delta G_t$ where $\Delta G_t(\Delta V_t, \Delta E_t)$ is the topological changes to the network graph G_t between time t and t_{t+1}. Nguyen et. al. considered four distinct event cases of these topological changes [18]:

- *new_node(u)*: A new node u with possibly new edges is introduced to the network.

- *remove_node(u)*: A node u and its neighboring edges are removed from the network.

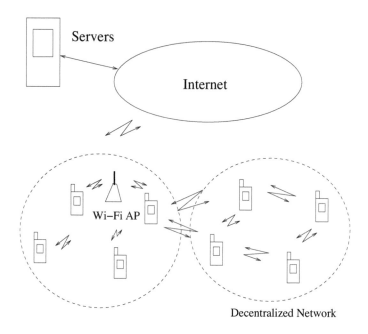

Figure 13.7: A hybrid MSN

■ *new_edge(e)*: A new edge *e* that connects a pair of nodes in the network is introduced.

■ *remove_edge(e)*: An existing edge *e* between two nodes is removed.

They proposed procedures to handle each of these events separately and an algorithm called *Quick Community Adaptation* (QCA) starts from the initial community structure C_0 and simply decides which procedure to call based on the event type received. Using QCA, they were able to detect communities in various MSNs such as the Facebook and the ENRON mail network [18].

A general model based on community life cycles was introduced in [12]. A dynamic network is represented as a set of time step graphs $G_1, ..., G_t$; $\mathcal{D} = \{D_1, .., D_k\}$ is the set of dynamic communities that may be present in one or more time steps, and the set of k_t step communities at time step t is identified by $\mathcal{C}_{\sqcup} = \{C_{t1}, .., C_{tk_t}\}$. Each dynamic community can then be represented by the history of its step communities. Based on this model, the key events are defined as follows:

■ *birth*: A new step community C_{ij} is observed at time t upon which a new dynamic community D_i containing C_{ij} is created.

■ *death*: If there is not a corresponding step community for a dynamic community D_i for at least d consecutive steps, D_i is removed from the set $\mathcal{D}$.

■ *merge*: If two dynamic communities D_i, D_j at time $t - 1$ match a single step community C_{ta} at time t, they start sharing a common timeline.

■ *split*: If a single dynamic community D_i at time $t - 1$ matches to two single step communities C_{ta} and C_{tb} at time t, a dynamic community D_j that shares the timeline with D_i until t, but has a distinct timeline after t is formed.

■ *expansion*: A single dynamic community D_i at time t has a significantly greater step community than the previous one.

■ *contraction*: A single dynamic community D_i at time t has a significantly smaller step community than the previous one.

The problem can then be formulated as a matching of step communities to dynamic set of communities at time t which can be viewed as a bipartite graph matching problem. Greene et al. proposed a heuristic threshold-based method to handle many-to-many mappings of the communities across different time steps, thereby enabling the identification of the merge and split events described above [12]. They applied the proposed method to a real mobile operator network uncovering a large number of dynamic communities with different evolutionary characteristics.

13.3.3 Middleware

Middleware in a computer system is used as the software layer between the application and the lower layers such as the protocol stack to provide common services needed by different applications. The middlewares for central and distributed MSN applications differ in terms of design and implementation issues. In central MSNs, the users transfer their current social data to the central servers which process this metadata and handle community formation [4]. The decentralized MSN networks, however, provide data transfer when this is physically possible. Figure 13.8 displays the contents of a typical middleware for MSNs which consist of upper and lower layers [4]. Management of social data by collecting and processing it; and management of communities along with social multicast and handling security can be directly used by the application whereas context management service can be reached by these higher level modules.

There is a number of requirements in the design of middleware for MSNs. Firstly, fully distributed middleware may be preferable as it allows users to access data efficiently without having a bottleneck central server. The battery power is limited for most of the mobile users such as the users of mobile phones. The middleware should therefore be energy efficient and also scalable so that system performance does not degrade drastically with the increase in the number of users. The middleware should also employ some kind of security mechanism to provide data privacy. Additionally, the users of the MSN have a range of mobile devices and the topology of the network is dynamic which makes the design of middleware for MSNs a challenging task. We will now describe sample middleware modules that are used in MSNs.

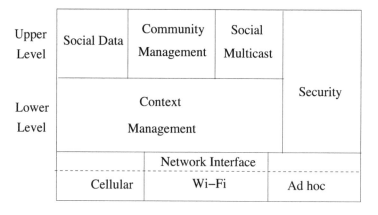

Figure 13.8: MSN layers

13.3.3.1 MobiSoC

The MobiSoC is a middleware built on the community concept [13] and it provides a common platform for capturing, managing and sharing the social states of communities. Profiles of persons, places and people-to-people affinities are the main components of a social community in MobiSoc. The social state of a community changes dynamically as new users, relations and events occur. MobiSoc has the people, the place and the location modules to collect data. The people module manages user data such as the identities, social interests whereas the place module deals with buildings and events associated with buildings and the location module is used for user location updates.

MobiSoc discovers people-to-people affinities by examining user profiles and various social factors. It uses a learning algorithm to find ad hoc clusters. The privacy management is handled by a central server which manages all user data reducing the processing burden of the client software running on the mobile user device. However, central servers are single points of failures and scalability is the major concern as in any other central processing system.

13.3.3.2 MobiClique

MobiClique middleware for MSNs allows persons to maintain and extend their online social networks using opportunistic encounters in real life [19]. When two mobile users meet by opportunistically and if their profiles have some common attributes such as friendship or interest, they are alerted and they can have an exchange which can be a friendship, content distribution or similar. MobiClick also proposes an MSN API that allows new applications to be built on top of it.

Each MobiClique node runs a periodic loop that consist of three steps: neighborhood discovery; user identification and authentication, and data exchange. Neighborhood discovery can be based on Bluetooth device discovery or broadcast beacons.

When a new device is discovered, identification phase starts where devices initiate a communication link to exchange user identity information. During the first encounter or when there is a change in profiles, full profiles are exchanged. After the identification step, data exchange between the users can be performed using multi-hop message communications which can be point-to-point or multicast.

Any application that needs to use the MobiClique first registers with it using the API which manages the local node social profile. Applications are informed by MobiClique about various events such as contact events or incoming messages. MobiClique is implemented using C++ and C# on a Windows mobile platform and Bluetooth was used for device discovery during initial tests. It was tested using the mobile social networks, asynchronous messaging and epidemic newsgroups applications successfully [19].

13.3.3.3   SAMOA

Socially Aware Mobile Architecture (SAMOA) enables the creation of anytime, anywhere semantic context aware social networks for MSNs [5]. It allows the users to create roaming social networks which reflect the close encounters around mobile user movements. SAMOA social networks are based on place and profile visibilities. The place visibility is the visibility of the user's physically visible place and the profile visibility of user's place or requirements and characteristics. The SAMOA social network management provides three management roles as follows:

- *Managers*: These are the users which create social groups and they are responsible for defining the scope boundaries of their social groups.

- *Clients*: These are the users located within the discovery scope boundaries of a social group and they can become the members of the group.

- *Members*: These are the users assigned to a social group.

A mobile user can assume any of the roles. In SAMOA, each manager of a social group is the center of a place and clients are connected to the manager by a routing path of a maximum of h hops called the *place radius*. Places may overlap such that users can be clients of two or more places at the same time. All SAMOA entities, whether places or users, have unique identifers and profiles to describe them. A place profile has an identification including an identifier, name and description of the place; and the activity part provides the social activities associated with the place. The user profile consists of *identification* and *preferences* parts. The identification part includes personal information about a user and the preference part provides the activities that the user is interested in. SAMOA employs a semantic matching algorithm to find clients related to the place profile activities. The second matching algorithm used selects users with attributes semantically matching the preferences of the place manager's discovery file as members. The SAMOA middleware consists of the basic service layer and social network management layer. The basic service layer is responsible for naming, detection of entities and point-to-point and multi-point communications based on UDP. The place dependent social network manager of the

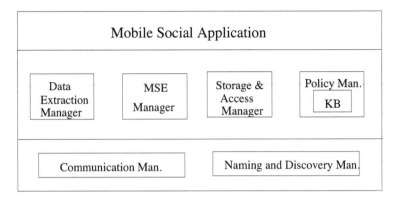

Figure 13.9: Yarta structure (adapted from [21])

middleware manages the membership of the social network whereas the global social network manager creates a global social network by integrating all place-dependent social networks.

13.3.3.4   Yarta

Yarta was proposed as a middleware support platform for mobile social applications [21]. It consists of two layers: the mobile social ecosystems (MSE) management middleware layer which manages and allows access to MSE data, and the mobile middleware layer to handle low level communication and synchronization. The MSE management middleware contains the data extraction manager, MSE manager, storage and access manager, and the policy manager. The mobile middleware includes the communication manager and the naming and discovery manager as shown in Figure 13.9.

Yarta uses resource description framework (RDF) which is a base semantic Web standard to represent the base data model. Data represented using this model are connected to form a uniform graph of social information which is managed by the knowledge base (KB) middleware component. The KB component provides APIs at different levels to access, update and remove social data as well as merging MSE graphs of different users. KB is enclosed in the policy manager module for access control according to the policy defined, as shown in Figure 13.9. The policy manager component provides definition, management, evaluation and enforcement of access control policies of KB. The storage and access manager provides the interface between the user applications and the KB by means of local high level queries and remote queries. Users can add/delete/query relations between the resources by using the storage and access manager.

Yarta is designed for ubiquitous and heterogeneous environments and its communication layer provides synchronous and asynchronous message passing using multi-radio links. It provides base security features using authentication and confidentiality methods. Yarta is developed in Java for portability and can execute on different user devices such as mobile phones and laptops. Its scalability and core features were tested on mobile phones with success [21].

13.4 Chapter Notes

We first reviewed the two fundamental types of ad hoc networks as MANETs and WSNs in this chapter. We then investigated clustering in these networks and described sample algorithms to perform clustering. Clustering in ad hoc networks is a well studied area of research as it has immediate practical applications. For example, a connected dominating set of a MANET or a WSN can be used as a backbone for routing purposes. Surveys of clustering in MANETs and WSNs can be found in [23] and [1].

We then described MSNs which are mobile networks technically and social networks conceptually. These networks have begun to emerge recently and there are a number of topics that can researched in these networks. One such area is the community detection which is more challenging than in static complex networks due to the mobility of the nodes and therefore should be performed in real time. The static community detection algorithms can be used to discover communities in MSNs to take consecutive snapshots of the network. However, new methods are needed since continuous execution of these algorithms are time costly. We have investigated one such approach where events that change the network are identified and procedures for corresponding events are executed when events occur. In another approach focusing on community life cycle, the changes in dynamic community structures due to step communities can be specified in terms of events. The problem can then be reduced to a special case of bipartite matching problem in graphs. Another challenge in MSN networks is the design and implementation of the middleware layer for centralized and decentralized MSNs. There are a number of such middleware modules in use for WSNs and we have described few recent studies. A detailed survey of MSNs is given in [15] and the surveys of middleware platforms for MSNs are provided in [4] and [14].

Clustering in MANETs and WSNs is a thoroughly studied area of research, however, community detection in MSNs is a relatively new research area with potential new findings. The design of middleware for MSNs has been realized in several systems that are in operation but each one of these systems has advantages and disadvantages. Thus, there is still a need for efficient middleware to accomodate needs of diverse user groups with different objectives. Designs of efficient algorithms to handle the task of community detection in MSNs and design of distributed, scalable and secure middleware to handle social data and community management are potential areas of research in MSNs.

Exercises

1. Discuss the real-life applications of MANETs and WSNs. Give examples of mobile sensor networks.

2. Compare the clustering algorithms in MANETs and WSNs in terms of their goals and the challenges faced.

3. Work out the clusters found using the *Lowest_ID* algorithm in the sample network of Figure 13.10 by showing the CHs and the gateway nodes.

4. Implement the highest degree algorithm to find the clusters in the sample network of Figure 13.11. Show all CHs and gateway nodes.

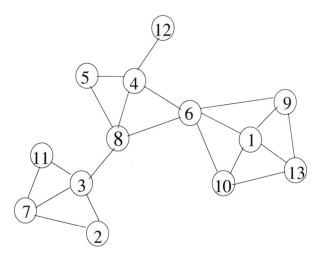

Figure 13.10: Example graph for Ex. 3

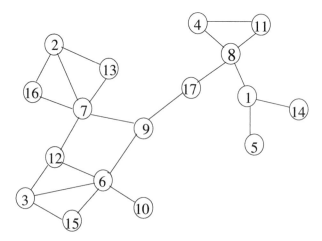

Figure 13.11: Example graph for Ex. 4

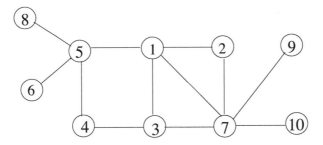

Figure 13.12: Example graph for Ex. 5

5. Show the two steps of the execution of *Wu_CDS* algorithm in the sample graph of Figure 13.12 to find a CDS of this graph.

6. Compare the middleware systems MobiClique, MobiSoC, SAMOA and Yarta for MSNs in terms of their architectures, application environments, advantages and disadvantages.

References

[1] A.A Abbasi and M. Younis. A survey on clustering algorithms for wireless sensor networks. *Computer Communications*, 30:14-15, 2826-2841, 2007.

[2] K.M. Alzoubi, P-J. Wan, and O. Frieder. New distributed algorithm for connected dominating set in wireleess ad hoc networks. In *Proceedings of 35th Hawaii International Conference on System Sciences*, pages 3849-3855, 2002.

[3] S. Banerjee and S. Khuller. A clustering scheme for hierarchical routing in wireless networks. Tech. Report CS-TR-4103, University of Maryland, College Park, 2000.

[4] P. Bellavista, R. Montanari, and S.K. Das. Mobile social networking middleware: A survey. *Pervasive and Mobile Computing*, 9:437-453, 2013.

[5] D. Bottazzi, R. Montanari, and A. Toninelli. Context-aware middleware for anytime, anywhere social networks. *IEEE Intelligent Systems*, 22(5):223-232, 2007.

[6] Y.P. Chen, A.L. Liestman and J. Liu. Clustering algorithms for ad hoc wireless networks. In *Ad Hoc and Sensor Networks*, (Eds. Y. Pan and Y. Xiao), Nova Science Publishers, 2004.

[7] D. Cokuslu, K. Erciyes and O. Dagdeviren. A dominating set based clustering algorithm for mobile ad hoc networks. In *International Conference on Computational Science*, Vol. 1, pages 571-578, 2006.

[8] Erciyes, K. *Distributed Graph Algorithms for Computer Networks*, Springer (Computer Communications and Networks Series), Chapter 15, 2013.

[9] K. Erciyes, D. Ozsoyeller, and O. Dagdeviren. Distributed algorithms to form cluster based spanning trees in wireless sensor networks. In *Proceedings of ICCS 2008*, Springer Verlag, 519-528, 2008.

[10] M. Gerla and J.T.C. Tsai. Multicluster, mobile, multimedia radio network. *Wireless Networks* 1:255-265, 1995.

[11] C.R. Lin and M. Gerla. Adaptive clustering for mobile wireless networks. *IEEE Journal on Selected Areas in Communications*, 15(1):1265-1275, 1997.

[12] D. Greene, D. Doyle, and P. Cunningham. Tracking the evolution of communities in dynamic social networks. In *Proceedings of International Conference on Advances in Social Networks Analysis and Mining*, pages 176-183, 2010.

[13] A. Gupta, A. Kalra, D. Boston, and C. Borcea, MobiSoC: a middleware for mobile social computing applications. *Mobile Networks and Applications*, 14(1):35-52, 2009.

[14] A. Karam and N. Mohamed. Middleware for mobile social networks: a survey. In *Proceedings of HICSS, 45th Hawaii Int. Conf. on System Sciences*, pages 1482-1490, 2012.

[15] N. Kayastha, D. Niyato, P. Wang and E. Hossain. Applications, architectures, and protocol design issues for mobile social networks: a survey. In *Proceedings of the IEEE*, 99(12):2130-2158, 2011.

[16] M. Newman. Fast algorithm for detecting community structure in networks. *Physical Review E 69*, 066133, 2004.

[17] M.E.J. Newman, M. Girvan. Finding and evaluating community structure in networks. *Physical Review E 69*, 026113, 2004.

[18] N.P. Nguyen, T.N. Dinh, X. Ying, and M.T. Thai. Adaptive algorithms for detecting community structure in dynamic social networks. In *Proceedings of INFOCOM 2011*, 2282-2290, 2011.

[19] A. Pietilainen, E. Oliver, J. LeBrun, G. Varghese, and C. Diot. MobiClique: middleware for mobile social networking. In *Second ACM Workshop on Online Social Networks* New York, pages 49-54, 2009.

[20] F. G. Nocetti, J. Solano-Gonzlez, and I. Stojmenovic. Connectivity-based k-hop clustering in wireless networks. *Telecommunication Systems*, 22(1-4):205-220, 2003.

[21] A. Toninelli, A. Pathak, and V. Issarny. Yarta: a middleware for managing mobile social ecosystems. In *Proceedings of 6th Int. Conf. on Advances in Grid and Pervasive Computing*, Springer, 6646:209-220, 2011.

[22] J. Wu and H. Li. On calculating connected dominating set for efficient routing in ad hoc wireless networks. In *Proceedings of the 3rd Intl Workshop on Discrete Algorithms and Methods for Mobile Computing and Commun.*, pages 7-14, 1999.

[23] J.Y. Yu and P.H.J. Chong. A survey of clustering schemes for mobile ad hoc networks. IEEE Communications Surveys. 7(1):32-48, 2005.

Index